Sampling for Analytical Purposes

Dr Pierre Gy at a sampling seminar in Paris

Sampling for Analytical Purposes

Pierre Gy
The Paris School of Physics and Chemistry

Translated by

A. G. Royle

JOHN WILEY & SONS

Chichester · New York · Weinheim · Brisbane · Singapore · Toronto

French language edition *L'Échantillonage des Lots de Matière en vue de leur analyse*
© Masson, Paris, 1996

Copyright © 1998 John Wiley & Sons Ltd,
Baffins Lane, Chichester,
West Sussex PO19 1UD, England

National   01243 779777
International (+44) 1243 779777
e-mail (for orders and customer service enquires): cs-books@wiley.co.uk
Visit our Home Page on http://www.wiley.co.uk
or http://www.wiley.com

Other Wiley Editorial Offices

John Wiley & Sons, Inc., 605 Third Avenue,
New York, NY 10158-0012, USA

WILEY-VCH Verlag GmbH, Pappelallee 3,
D-69469 Weinheim, Germany

Jacaranda Wiley Ltd, 33 Park Road, Milton,
Queensland 4064, Australia

John Wiley & Sons (Asia) Pte Ltd, Clementi Loop #02-01,
Jin Xing Distripark, Singapore 129809

John Wiley & Sons (Canada) Ltd, 22 Worcester Road,
Rexdale, Ontario M9W 1L1, Canada

Library of Congress Cataloging-in-publication Data

Gy, Pierre.
  [Echantillonage des lots de matière en vue de leur analyse.
  English]
  Sampling for analytical purposes / Pierre Gy : translated by A. G.
Royle
      p.   cm.
  Includes bibliographical references (p.   ) and index.
  ISBN 0-471-97956-2 (alk. paper)
  1. Granular materials—Sampling. 2. Bulk solids—Sampling.
  3. Sampling (Statistics)   I. Title.
  TA418.78.G878   1998
  543'.02—dc21   97-49071
                                                          CIP

British Library Cataloguing in Publication Data

A catalogue record for this book is available from the British Library

ISBN 0 471 97956 2

Typeset in 10/12pt Times by Keytec Typesetting Ltd, Bridport, Dorset.
Printed and bound in Great Britain by Biddles Ltd, Guildford, Surrey
This book is printed on acid-free paper responsibly manufactured from sustainable forestry,
in which at least two trees are planted for each one used for paper production

At the end of half a century of research I dedicate this book to the Ecole Supérieure de Physique et Chimie Industrielles de Paris (ESPCI—Municipal School of Physics and Industrial Chemistry of Paris)—where I first learned to learn—and in particular to its Director, Pierre-Gilles de Gennes, winner of the Nobel Prize in Physics. I hope they will agree to being the trustees of a science that did not exist fifty years ago, which is still almost never taught, and which, without their support, runs the risk of sinking into oblivion altogether.

'To persuade people to act you must urge, urge and continue to urge'

Pierre-Gilles de Gennes
Jacques Badoz

Les Objects Fragiles

The author has been urging analysts for nearly 45 years, sadly without much success, at least in France. But he has no intention of giving up so quickly and this book is the proof of it!

> 'Geography was not listed among the "seven liberal arts" of the Middle Ages. For a thousand years, [...], its status unrecognised, geography remained an orphan in the world of knowledge. It was an incredible jumble of ideas, some true, some imaginary, [...] of biblical dogma, [...] of mythical fantasies.'

Daniel Boorstin

The Discoverers

As for waiting a thousand years for status to be awarded to the sampling orphan, that is for analysts to decide after reading this book. In its pages will be found examples, applied to sampling, of the 'incredible jumble' Boorstin spoke of: dogmatic assertions and expressions as empty as they are abrupt.

And from Boileau (*Le Lutrin*, 1674–83):

> 'Work at your own pace, never be rushed,
> Into mad speeds never be pushed,
> Make haste slowly whatever the pain,
> Polish unceasing, then do it again.
> Re-edit twenty times if needs to improve it,
> Add as you think fit, but often remove it.'

And finally, from Jean d'Ormesson (*Une autre histoire de la litterature française*):

> 'Montaigne is a one-book author [*Essays*, written between 1580 and 1592] ... he never stopped refining it, correcting it and adding to it.'

The last two quotations seem to apply to the author, who initiated his sampling theory nearly fifty years ago.

Contents

Preface to French Edition

This book is a much-abridged version of another entitled *Sampling of Heterogeneous and Dynamic Material Systems* published by Elsevier, Amsterdam, 1992 (reference [7]). The present book is lighter, easier to read and consult and more easily accessible to a wider public that includes academics, students and practising analysts.

Sampling lies in the realm of mathematical statistics but the teaching of statistics is currently restricted to only one of the three branches described in Appendix 2. To help the reader to understand the present approach and its nomenclature, several concepts with which he or she may not be familiar must be clarified at the outset.

THE CONCEPTS OF OBJECTS, CONSTITUENT ELEMENTS, SETS, POPULATIONS AND SERIES

Statistics deal with assemblages of 'objects' (in the broadest sense) which can be either material, such as the fragments comprising a lot of broken material, or immaterial, such as analytical results obtained from a particular sample. From a mathematical standpoint these assemblages of objects are called 'sets'. We are dealing here with two kinds of sets: 'populations' and 'series':

1. In a 'population' the objects are not presumed to be ordered. If some order does exist it is not taken into account.
2. In a 'series' the objects are ordered, usually on a geometrical or chronological basis. In the latter case what are called 'time series' may be referred to.

The 'constituent elements' making up a population of material objects may have the same physical mass, for example in the case of closely sized fragments, and can be given the same 'statistical weight'. When they have different masses, as in the case of undifferentiated fragments, they must be given 'statistical weights in proportion to their masses'.

Classical statistics, taught ubiquitously, apply only:

- To populations of non-ordered objects
- To objects having the same mass and therefore the same statistical weight.

It is a mathematical mistake, and a grave one in practice, to apply its laws:

- To sets (populations or series) of objects having different masses
- To a series of ordered objects irrespective of their masses

Appendix 2 gives a schematic illustration of the three branches of statistics:

- Statistics of populations of objects with the same weight (i.e. 'classical statistics')
- Statistics of populations of objects with different weights
- Statistics of a series of objects, i.e. chronostatistics.

Only the first of these is taught regularly throughout the world, despite the importance of the other two. The theories developed by the author are entirely original. A complete work of reference had therefore to be published first. It has been presented in French [6] and in English [7], in both of which the full mathematical treatment of the theory was presented to the scientific community for them to judge its validity.

Ten years have passed since reference [6] and six since reference [7] were published. The reviews of the English version [7] have been numerous and, on the whole, positive. None of them questioned in any way the methods by which the author considers sampling theory should be developed. This gives him reason to believe that the mathematical development of the 'generalised sampling theory', which is summarised in this book, has been accepted by the scientific community. This is why the mathematical proofs, assumed to be known or accepted, have been omitted, as have some chapters of a purely statistical nature that added to, but made heavy going of, the 1992 edition [7].

The sampling problem can be summarised in one simple question:

What must be done—or must be avoided—to obtain a 'representative' sample from some given lot?

From which there immediately follows a second question:

What, therefore, is a 'representative' sample?

A provisional answer at this stage, anticipating later developments, is that a sample is 'representative' when it has both the characteristic of 'accuracy' or 'unbiasedness', (the result of the qualitative approach) and the characteristic of 'reproducibility' (the object of the quantitative approach). Scientific definitions of these expressions are given in Chapter 4.

P.G.

Preface to English Edition

Nearly sixty years have passed since I was first instructed on how to take a representative sample. The material was charcoal carried in sacks by rail waggons, journeys occasioning much jigging of fines and other debris to the bottoms of the sacks (and sometimes put there by human hand too!). My instructions were to take a shovelful from the material at the top of every tenth sack. The resulting lot was then to be reduced in size and mass by coning and quartering, size reduction being achieved by smashing the material with a No.9 shovel on a concrete floor.

The reduced mass was finally ground in a disc pulveriser and screened, the fines going for analysis and the oversize being discarded. Such was the final sample I analysed, the results being recorded in percentages to the second decimal place, or one part in 10 000.

I was told this was the correct way to prepare the sample. Looking back I had, in short, been shown a rhinoceros and told it was a duck. However, since those days I have had many opportunities to examine other sampling and preparation procedures and to confirm that the rhinoceroses and ducks are still with us.

Everyone concerned with the taking and preparation of correct samples is, or ought to be, indebted to Dr Gy. Virtually single-handedly he has wrenched the so-called art of sampling away from the world of mumbo-jumbo, an example of which I have described above, and put it on a rational, scientific basis. His whole working life has been devoted to this end; we owe him much.

He has every right to complain of the lack of attention paid to correct sampling, observing that the commonest approach is to 'take what comes to hand, provided it doesn't cost too much'. Or, as I have heard the 'art' of sampling described, 'what goes in, goes in', an expression equivalent to that of Dr Gy's and a sure-fire way of collecting anything but a correct sample.

The truly surprising thing about incorrect sampling is that its incorrectness is so obvious, at least once it has been pointed out, and how so easily it could have been avoided. In the great majority of cases observing a few simple rules is enough to give correct samples, as this book so ably demonstrates.

Language translations have some of the characteristics of restoration work on a painting or tapestry. They can be overdone, resulting in something poorer than the original. I have tried to retain the flavour and vigour of Dr Gy's original text as far as I can, and any awkardnesses or infelicities in the English version are my responsibility.

Finally, I thank Dr Gy once again for giving me an opportunity to express my gratitude to him for the unique contributions he has made to the science of sampling.

A.G.R.

Acknowledgements

The translation of a scientific work requires at least three qualities: a knowledge of the two languages, an easy enough requirement to fulfil in the case of English and French, but more importantly it calls for a deep, professional knowledge of the subject involved, and in the case of sampling this is a great deal rarer.

Symposia were organised by the Institution of Mining and Metallurgy in London in January and September 1972 and July 1973 with the title 'Geological, Mining and Metallurgical Sampling', to which both the author and A. G. Royle were invited to present papers. As a result, these two participants recognised a common interest in the problems posed by sampling in the mining, mineral and metallurgical industries. Over the following years this recognition blossomed into a friendly correspondence that frequently overflowed the narrow confines of sampling and into that of History, with a capital letter, in particular the period between the allied landings in Normandy on the 6 June 1944 and the end of the war, and even more particularly into the liberation of Paris on 24 August 1944.

When the author wanted to have an English translation made of the book he had written in 1996, for obvious reasons his first choice for this task was Bon Royle. Bon accepted with good grace what must have been a heavy load for him to shoulder, but the results are here and everyone may judge their quality. The author wishes to offer Bon Royle the deep gratitude of a sincere and grateful friend.

General Plan of This Book

The book is divided into an introduction, four parts and three appendices:

Chapter 1: Introduction

Part I: The qualitative approach (Chapters 2 to 7). Ensuring 'accuracy' when sampling 'sets' of material elements, whether of populations or of series.

Part II: The quantitative approach (Chapters 8 to 10). Ensuring 'reproducibility' when sampling 'populations', defined as sets whose elements are not ordered and which may or may not be correlated with one another.

Part III: The quantitative approach (Chapters 11 and 12). Ensuring 'reproducibility' when sampling 'chronological or geometrical series', defined as sets whose elements are ordered, and which may or may not be correlated with one another.

Part IV: Applications to problems other than sampling proper
The estimation of mass by 'proportional sampling' (Chapter 13).

Appendix 1: Brief résumé of classical statistics (non-weighted).
Appendix 2: Non-weighted statistics; weighted statistics; chronostatistics.
Appendix 3: Integral calculus nomenclature.

Chapter 1

Introduction

1.1 THE PARADOX OF THE DIAMOND AND THE SAMPLE

It is a paradox that there are infallible physical tests that enable a real diamond to be distinguished from a valueless paste fake, whereas there is no method that enables a truly 'representative sample' to be distinguished from a 'specimen' which, because of the way it was collected, cannot be regarded as a reliable 'sample'. If no details of the way in which it was taken are known, neither the sample nor the specimen gives the slightest hint of what should be its true function—to show whether it is representative, or not representative, of what it purports to represent. This is one of the reasons why the science of sampling is subtle and confusing to the point where its usefulness, even its very existence, slips unnoticed past certain minds that are otherwise quite brilliant, particularly those of analytical chemists.

1.2 THE ROLE OF SAMPLING IN VALUATION

Sampling is an auxiliary process in valuation, and three cases can be recognised:

1. Valuation can be done directly

   - On the whole of the object to be valued and
   - In a way that is sufficiently accurate and reproducible. It is the only case—and a very rare one—in which sampling is not needed.

2. Valuation can be done directly

   - On the whole of the object to be valued but
   - In a way insufficiently accurate and reproducible.

This is particularly the case when continuous measurements are made of the masses or volumes of moving streams of material. They can be assessed by proportional sampling, a very precise but unfortunately undervalued method which has nevertheless proved its reliability in industry and in pilot plants (see Chapter 13 below and Chapters 29 of references [6] and [7]).

3. Valuation cannot be done by taking the whole of what is to be valued.

This is generally the case with assaying, whether the subsequent analysis is by chemical or other means. Valuation has to be done on a fraction of the lot: the sample. For this operation to make sense the sample must be representative, a quality to be defined in Chapter 4. Sampling is thus necessary or recommended in the following cases:

- Chemical analysis
- The estimation of masses and volumes by proportional sampling

The sample submitted for analysis is more often than not small in mass, or is even the lowest mass both in absolute terms and in proportion to the mass of the object it purports to represent.

1.3   THE ROLE OF SAMPLING IN RESEARCH

In research, comparative tests are often made on a certain number of fractions taken from the same lot. The fractions are called 'twins' (even if there are more than two) when they are taken under identical conditions, apart from small random differences. The twinned fractions are in fact twin samples and as such they obey the laws of the theory of sampling

But, like their human counterparts, twins are never strictly identical. The errors causing the differences create an experimental scatter which, in whatever preparation or other routine follows (for example, if an analysis of variance is done), increases the residual variance; this reduces the efficacy of the tests and increases the number and hence the cost of the tests needed to reach a sure conclusion.

1.4   SAMPLING: AN UNAVOIDABLE PRECURSOR TO
ANALYSIS

It is impossible to analyse the whole of a lot whose composition is to be determined because:

- To analyse such a mass of material would cost too much and
- Analysis is often a destructive process.

Thus, the analytical process requires that the mass of the lot to be estimated is first reduced to the tiny amount finally taken for analysis.

1. Sampling in its strict sense is therefore a simple mass reduction. Mass

reductions preceding analysis must retain the composition of the original lot as closely as possible. Therein lies the kernel of the sampling problem. The laws of classical statistics apply only to sampling populations whose members have the same statistical weight, for example manufactured objects and the beliefs or intentions of human beings gathered in opinion polls. In these cases sampling is a simple reduction in their number.

In 1951 [1] the author wrote, for the company employing him at that time, an internal report containing an original sampling theory to cover populations of objects having different physical masses and hence different statistical weights, namely a set of mineral fragments. This theory, still as valid today, connected a certain component of the sampling variance (the 'fundamental' variance) to the properties of the lot and the properties of the sample. It was made public knowledge at an international congress in 1953 [2]. But as Arthur Koestler wrote, with good reason,

> As with contagious diseases, new ideas need long incubation periods before their effects are observed (*The Sleepwalkers*).

2. Sampling in its broad sense covers all the operations which, beginning with the object to be valued, the 'lot', end with the fraction of pulp that is finally analysed in its entirety. Thus, mass reductions (sampling in its strict sense) alternate with various 'preparation' stages such as transferring from point A to point B, crushing, pulverising, drying, homogenising, etc. All these operations generate errors that need to be detected, estimated and, if possible, eliminated.

The author regrets having to record that the generalised sampling theory, i.e. the sampling of sets (populations or series) of objects of different statistical weight, is confined by the scientific community to a kind of academic no-man's land, an intellectual ghetto having little or no contact with the outside world. It is as though analysis is considered to be an end in itself, rather than being one of the two inseparable phases of valuation.

Is this situation changing? Slowly, perhaps. In 1989, Etienne Roth asked the author to write the chapter on 'Sampling' in *Les Techniques de l'Ingenieur* [4] and stated unequivocally in his foreword [3]:

> The present article [...] deals with one of the most important problems facing those who have to interpret analytical results: their representativeness.
>
> Consequently, at every level concerned with the results of analyses, from the board of directors down to the technician making the analyses, there is a need to be aware that the way in which the samples were taken and prepared is just as important as the way in which they were analysed.
>
> This awareness leads to other things:
> - teachers of analytical methods, and quality control, must allow sufficient time for them to be taught properly. This is not always the case;
> - those who establish analytical laboratories must provide both the proper equipment and the qualified staff needed to produce representative samples;
>
> In short, analysts must be the first to refuse to pass on results that are supposed

to represent a lot if they have not the necessary guarantees about the samples that have been offered to them.

Professor Wolfhard Wegscheider of the University of Graz (Austria) recently published (1993) under the title 'Richtige Probenahme: Voraussetzung fur richtige Analysen' (Accurate sampling: a preliminary to accurate analyses) an article [5] in which he refers mostly to the theories set out in references [6] and [7].

Professor Erno Pretsch of Zurich Polytechnique wrote in 1995 under the title 'The forgotten bias' [8] an editorial that served as an introduction to an article the author wrote for the same issue of *TrAC* [9]:

> Who is responsible for sampling errors? In most cases it is simply assumed that this only concerns 'others'. Does the analyst feel that his responsibility begins with receiving the sample to be analyzed? Does he care whether it is just a 'specimen' (i.e. non-representative of the batch) or a real sample? In general he focuses on improving his technique in view of good repeatability, reproducibility, and robustness and does not realise that all this makes no sense if the sample is biased. But who else should take the responsibility? Unfortunately also in official regulations care is mainly attributed to obtaining high precision and much less to accuracy. With an increasing sensitivity of analytical methods, both in terms of required sample amount and concentration, sampling problems become more and more serious.

Four conclusions can be drawn from this section:

- No analysis is possible without first preparing a sample.
- The analyst's work makes no sense if the sample is biased.
- Few analysts know that a sampling theory exists.
- The teaching of analysis must include sampling, as no one else is likely to teach it except teachers of analytical chemistry.

1.5   THE PARTIES INVOLVED IN SAMPLING

1. *Researchers in sampling*: They are welcome if they can advance the theories developed in the author's previous work. The reader should watch out particularly for papers to be published shortly by Dominique François-Bongarçon (Mineral Resources Development Inc., San Mateo, California) and by Geoffrey J. Lyman (Julius Kruttschnitt Mineral Research Centre, University of Queensland, Australia), the latter in the *International Journal of Mineral Processing*, especially as both are rather critical.
2. *Those who use the results*: They are interested above all else in the reliability of the results and not in the degree of sophistication of the equipment. Consequently, they have no choice but to apply the rules of the generalised sampling theory.

3. *Industrial, technical and commercial undertakings*: Need to be aware of the dangers of sampling when they treat it simplistically as a materials-handling problem and not as a science. (Philosophy: take whatever comes to hand provided it doesn't cost too much!)

4. *University departments of analytical chemistry*: Must understand that analysis is not an end in itself, and they need to ensure they are analysing properly taken, reliable samples. They must also teach the theory of sampling because no one else is likely to if they don't.

5. *Standards committees*: Are content to say that samples must be 'representative' without ever defining the word objectively. They should seek scientific advice, but when questioned on this point reply that it is not their concern. The business of legislating on standards is left to technical committees, all too often comprising agents drawn from administration or commerce, who are more concerned with guarding their interests than with science (there are some notable exceptions, such as ISO-TC 183). Membership of these committees is voluntary and decisons are arrived at on the democratic principle of majority voting, a very dangerous practice where scientific matters are involved.

6. *Equipment manufacturers*: The theory presented here was first developed to solve the problems of the mining and metallurgical industries. With the gradual contraction of the European mining industry the sampling equipment manufacturers who followed these theoretical developments are no longer in business. There remain only the makers of material-collecting machines, some of them mechanically sophisticated (which inspires some confidence) but functionally inadequate as they do not follow the rules set out by the theory. It is forgotten that sampling is a complete science and not a simple gathering technique for taking at the lowest cost whatever parts of the lot are most easily reached by the sampling tool.

The same problem faces the makers of analytical equipment, who have yet to realise that returning an analytical result to three or four decimal places is pointless when the sampling error has every chance of being bigger than the first of them. There is on record the case of the graduate technician assaying gold with an atomic absorption mass spectrograph. As in the case of M. Jourdain and his prose (he spoke it unknowingly (*Le Bourgeois Gentilhomme*, Molière)), the technician prepared her samples in similar fashion and thus failed to realise she was making a systematic sampling error of the order of . . . 50%. Nothing in her previous training, nothing in the maker's manual, had warned her against making a sampling error she could so easily have avoided.

However, it may be possible to release the manufacturers from their culpability, at least partially. Their aim is to make and sell equipment. If their industrial clients, the universities and the standards committees make no demands on them for scientifically designed equipment, it is hard to see why they should put themselves to the trouble of consulting specialists.

1.6   SAMPLING ERRORS AND THE RISKS INCURRED

1. *Risks involving the lives and health of human beings*: In biology, in medicine, in the pharmaceutical industries, in the food industries, in the control of water supplies and effluents, etc. the health and lives of human beings depend on the results of analyses. The regrettable mistakes made in hospitals from time to time, even in highly developed countries, which are usually attributed to analytical errors are more often probably due to sampling errors. Are the teachers in medical schools aware of this possibility?
2. *Financial risks*: Almost all human activities, such as production, manufacture, trade in primary and finished products, legal and customs controls, supervising the release of harmful products and industrial waste into the environment, all need to be monitored by sampling and analysis.

When human lives are at risk the stakes rise above mere financial considerations. In industry, sampling errors can incur severe financial penalties. The sums involved can amount to millions or even billions of dollars.

In one legal case between a tin mine and a smelter it was shown that biased sampling at the smelter had undervalued the mine's concentrates, over a period of several years, by a factor of 9%. What industry would knowingly allow 9% of the value of its production to be discounted? Is there any other operation capable of leading to such a loss? Never!

1.7   EXAMPLES OF COSTLY SAMPLING ERRORS

The following examples are taken from the author's experience as a consulting engineer and expert at the Court of Arbitration of the International Chamber of Commerce:

1. *A concentrator built on the basis of biased samples*: An unsuitable mill (designed and built in North America) had to be dismantled. The loss, in 1960, was of the order of $10 million.
2. *Exploitation of mining blocks*: This took place in 1982 in Australasia, at what was then the world's second biggest copper mine. Extremely biased samples were taken by hand from cones of blast-hole cuttings, introducing 2% of waste dilution and bringing about an annual loss of some $8 million.
3. *Mineral sales*: During the course of a contract between a South American tin mine and a European smelter, biased manual sampling (in the case referred to above) cost the mine $7 million over three years.
4. *The extra cost of correct equipment*: The term 'correct sample' will be defined in Chapter 4. For the time being it is enough to say that it is the best sample that one knows how to take in practice. While it may be true that correct equipment sometimes costs more than incorrect equipment of the kind that uses manual sampling, experience shows that the extra cost is always

recovered quickly (in weeks or months). It is a question of choosing between the long and the short term. Industrialists aware of economic realities must always plan for the long term.

5. Unfortunately, incorrect sampling equipment capable of generating biases of 5–20% can be found in plants and described in the catalogues of machinery manufacturers.

6. Analytical equipment manufacturers often suggest that secondary sampling devices are incorporated into their appliances. These rarely pay any regard to sampling theory.

1.8 THE COMPLEMENTARY NATURE OF SAMPLING AND ANALYSIS

1.8.1 Analysis of the Process by which Quality is Estimated

This is illustrated in Figure 1.1.

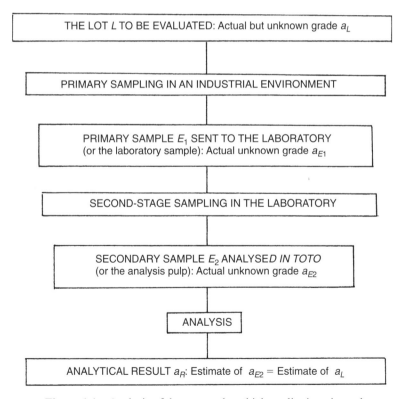

Figure 1.1   Analysis of the process by which quality is estimated

1.8.2 Responsibility and Competence

These can be allocated as follows:

PRIMARY SAMPLING IN THE INDUSTRIAL ENVIRONMENT

Situation: Industrial, outside the laboratory
Responsibility: Often undefined, rarely the analyst's
Qualifications: Usually non-existent

SECONDARY SAMPLING IN THE LABORATORY

Situation: In the laboratory
Responsibility: The analyst's
Qualifications: Usually non-existent

ANALYSIS

Situation: In the laboratory
Responsibility: The analyst's
Qualifications: Nearly always excellent

1.8.3 Additivity of Sampling and Analytical Errors

The global estimation error GE can be analysed as follows:

TOTAL ERROR AT THE PRIMARY SAMPLING STAGE TE_1

$$TE_1 = (a_{E1} - a_L)/a_L$$

TOTAL ERROR AT THE SECONDARY SAMPLING STAGE TE_2

$$TE_2 = (a_{E2} - a_{E1})/a_L$$

TOTAL ANALYTICAL ERROR AE

$$AE = (a_R - a_{E2})/a_L$$

GLOBAL ESTIMATION ERROR GE

$$GE = (a_R - a_L)/a_L = TE_1 + TE_2 + AE$$

GE is the sum of three components with independent laws of probability.

1.8.4 Consequences of the Additivity and the Independence of Probabilities

ADDITIVITY OF MEANS (BIASES)

$$m(GE) = m(TE_1) + m(TE_2) + m(AE)$$

The sampling bias can be far greater than the analytical bias (more than 1000 times).

ADDITIVITY OF VARIANCES

$$s^2(GE) = s^2(TE_1) + s^2(TE_2) + s^2(AE)$$

The sampling variance can be considerably greater than the analytical variance.

CONSEQUENCES OF THE ADDITIVITY OF BIASES AND VARIANCES

The analyst's work makes no sense if the sample is biased (Erno Pretsch) [8]

As much care and cash must be given to sampling as to analysis (Etienne Roth) [3]

1.8.5 Materials Containing Trace Elements

With the development of analytical methods for concentrations as low as ppb (10^{-9}), ppt (10^{-12}), ppq (10^{-15}), to say nothing of the detection of individual molecules [11] and [12], it is clear that the analyst, with his mind focused exclusively on the analytical problems facing him, is light-years away from solving the sampling problems of which he is unfortunately all too unaware. For him, the analysis is all.

The theory shows that the lower the concentration of the element, the larger the mass of sample that needs to be taken to achieve a given reproducibility, and the greater the error made by ignoring the theory or by using invalid statistical formulas.

In the case of trace elements sampling errors can affect the analytical results in terms measured in orders of magnitude. Up to the present this point has been completely ignored by universities and standards committees.

It is time that interested parties took note of this state of affairs.

1.9 THE RELATIVE IMPORTANCE GIVEN TO SAMPLING AND TO ANALYSIS

The biases observed in practice can be as high as 1000% at the primary sampling stage and 50% at the secondary stage, whereas the analytical bias rarely exceeds 0.1–1% (a little higher for trace elements). World-wide, investment in reliable equipment is in a ratio estimated to be of the order of 1000 to 1 in favour of analysis. Analysts are ready to buy the latest piece of analytical equipment provided it looks impressive enough, but they are not prepared to pay as much for a sampling system, reliable but not so spectacular, that respects a theory they choose to ignore.

Sampling is not fashionable! Above all, it is possible to take excellent samples without the aid of the indispensable computer without which, in these times it seems, no equipment can be taken seriously. (*Note*: The author uses a computer as necessary, but does not offer this as a proof of his intellectual abilities.)

Published papers are in a ratio of some 100 to 1 in favour of the analyst. Moreover, some of the sampling papers are somewhat dated [13]. The teaching of the generalised sampling theory is non-existent, at least in Europe. That of analysis is ubiquitous, excellent and exhaustive. Sampling standards (ISO among others) take no notice at all of the existence of the generalised sampling theory.

1.10 A CARICATURIST'S LOOK AT THE BIBLIOGRAPHY OF SAMPLING

Before embarking on considerations of a theoretical nature, and deliberately setting to one side what the author has written on the subject since 1951, it may be as well to look at what has been written, some good, some less than good, on sampling.

- 1930: Grummel and Dunningham join Arthur Koestler in saying:

  Those whose interest in sampling is recent will find it difficult to understand how hard it is for authors to have new ideas adopted.

  As with contagious diseases, new ideas require a long incubation period before their effects are recognised. (Koestler)

- 1967: Kaye (Illinois Institute of Technology): 'Sampling, the Cinderella of the analyst's art' (Rock Products). A fair observation, this, worthy of being framed:

  The accuracy of many analytical data reports is a mirage because unwitting negligence and false cost consciousness have ensured that a sample of powder taken with cursory swiftness has been examined with costly precision.

- 1968: *Dictionary of Mining, Mineral and Related Terms* (US Bureau of Mines): There is a splendid example of cant under the section headed 'Sampling', i.e.:

  Honest sampling requires good judgement and practical experience.

- 1981: K. and T. (Analytical Chemistry). An anthology from the same press as the last.

  Random sampling is difficult (p. 925A).

  ... Obviously, a representative sample cannot be selected by a random process (p. 926A).

Such obviousness should be distrusted. The statement is false, as will be demonstrated later.

  A good approach is to collect a small number of samples, using experience and intuition as a guide to making these as representative of the population as possible (p. 930A).

  Sampling theory cannot replace experience and common sense (p. 938A).

When one wishes to ignore a theory one doesn't understand, a common practice is to call on 'experience, intuition and common sense'. It is the eternal argument between the theoreticians and the experimentalists. It makes more sense to turn the sentence round and say 'experience and common sense cannot in any way replace the theory', although they are useful and can be included in the development of a theory.

  Sampling is not simple (p. 938A).

The author, having spent over 45 years developing the subject, shares their opinion.

- 1990–95: Standards:

All the analytical standards insist that 'analyses must be carried out on representative samples'.

As far as the author is aware, no scientific definition of a 'representative sample' is given in any of these publications, which do not say what should or should not be done in order to obtain one. However, a definition has been given in all the author's publications since 1975, as well as the recommendations he describes for taking representative samples, i.e. samples that are both accurate and reproducible (see Chapter 4 and references [6] and [7]). In addition, and this is more worrying, the recommendations made by some sampling standards are either arbitrary or contrary to the theoretical conclusions, and often both. While not wishing to cause aggravation, it has to be said that any use made of the standards must be done with extreme circumspection.

According to a highly placed scientific person in the ISO, the role of the Technical Committees (ISO-TC) is limited to 'describing the practices on which commerce has based itself for a long time'. If this really is the general philosophy of the standards committees, one can only disagree with it in the strongest terms. Standards committees concerned with scientific subjects such as sampling should seek support in a recognised theory, not contradict it.

- 1995: Laboratory of the Government Chemist [13] (Section 2.2.1): 'Representative sample: this is a sample that is typical of the lot'. But what is a 'typical sample'?

1.11   THE FUTURE OF ANALYSIS

The future of the science of analysis, i.e. improvements in the sensitivity, accuracy and reproducibility of analyses, is limited:

- Not by the quality of the equipment or by the ability of analysts who attain or approach perfection but
- Solely by a failure to submit representative samples to analysis, particularly in trace or ultra-trace concentrations.

Quality estimation is a chain and sampling is its weakest link.

To remedy this state of affairs, i.e. to advance the science of analysis, the general theory of sampling must be taught in universities and understood by all affected parties.

1.12   BEGINNINGS OF A GENERAL SAMPLING THEORY

When in 1949 the question of sampling minerals was first put to the author he realised that the essence of the problem lay in the need to 'control', in the fullest sense, sampling errors. That is, to eliminate them wherever possible, or otherwise to minimise them and then to estimate them. The bibliography of sampling shows that historically two lines of approach have run concurrently.

1.12.1   An Experimental Approach

This is more widespread, and involved: Beginning with a given lot, to collect a certain number N of samples, and prepare and analyse them in the same way (values $a_1, \ldots a_n, \ldots, a_N$)

- Calculating a mean $m(a_1)$ and a variance $s^2(a_1)$
- Altering the conditions of the experiment, e.g. the maximum size d of the fragments, the mass of sample M_E, the grade of the material to be sampled, etc. and then
- Trying to establish algebraic formulas connecting the conditions and the results of sampling, i.e. between the value of each relevant variable (primarily the size d of the largest fragments and the mass of the sample M_E) and the variance $s^2(a_1)$. The mean $m(a_1)$ was usually passed over in silence despite its importance—it does, after all, determine the accuracy of the operation.

This has been the route followed since about 1930 by many researchers in a once-flourishing coal industry. It continues to be recommended by a number of authors and standards committees. But as Albert Einstein, a master of his subject, said: 'A theory can be verified by experiment, but there is no path leading from experimentation to the development of a theory'

Formulas derived from experiments are essentially unable to assist in the development of a theory that leads to a full understanding of the phenomenon under study. Experiments cannot replace logical reasoning.

1.12.2 A Purely Logical Approach

This is encapsulated in the word 'theory'. What will be presented here is an attempt to explain, by analysing them, all the factors giving rise to sampling errors. A theory is neither an argument nor a hypothesis; it is a logical or mathematical walk whose every step leads to an ineluctable proof. An approach of this nature was made a century ago by Brunton in a study [10] based on geometric similarity. According to him the mass M_E of a sample had to be proportional to the cube of the size d of its largest fragments.

It has been decided therefore to keep to a logical path, to put the accent on the word 'understand'. In this book the author has tried, above all else, to lead his readers to an understanding of the processes by which sampling errors are generated.

1.12.3 Validity of the Proposed Theory

The theory was first developed from studies of solid fragments of mineral origin, such as run of mine ore, concentrates, mill tailings, the raw materials and finished products of the cement industry, etc. However, in the mathematical model that was developed during these studies there is nothing that requires a grain of vegetable origin to be distinguished from one of mineral origin, or from any other discrete particle from whatever source. In fact the theory has been used to sample cereals, sugarbeet, and even deliveries of cattle bones imported from Asia to make gelatine.

In the same way, with a change of scale, what is true for solid fragments is equally true for the molecules and ions of liquids. The results derived for solid particles are equally valid for all the liquids and multi-phase mixtures found in the chemical, pharmaceutical and food industries. Likewise, the theory has biological applications to all bodily fluids such as blood and urine, although it must be acknowledged that this branch of science is totally unaware that a sampling theory exists.

The protection of the environment is a major preoccupation in these times. The theory can be applied to domestic waste, to industrial effluents, to any material discharged into the environment, and to the supply of clean water. It applies equally to the detection of fraud and to customs control.

1.13   CONCLUSIONS

It appears indisputable (and see Appendix 1):

- That all concerned become aware that classical statistics apply only to the sampling of populations of non-ordered objects of the same, or nearly the same, mass, or which have the same statistical weight; the possibility of correlations between these objects is not taken into account
- That the statistics of populations of objects having the same mass is not valid for sampling populations of objects of different masses
- That the statistics of populations of objects whose masses may be equal or unequal do not apply to time- or space-dependent series, such as a series of increments taken from a moving belt (very important in industry) in which correlations are of the greatest importance and must be taken into account by the model used to characterise the system
- That everyone becomes aware of the additivity of errors, of biases, and of the variances of sampling and analysis
- That the general theory of sampling be taught in universities at two levels: one for the engineer, another for the higher-grade technician, preferably in analytical chemistry departments which seem to be the only ones sufficiently motivated to do so
- That, as counselled by Etienne Roth [3], industry and researchers must agree to investing in correct sampling equipment
- That, again following Roth, analysts must refuse to pass on their results if they do not have the necessary guarantees about the representativeness of their samples
- That standards committees finally stop viewing sampling as a simplistic material-gathering operation whose sole aim is to take the most easily reached parts of a lot, without reference to any considerations of a theoretical nature
- That analysts finally stop regarding analysis as the only factor in providing an accurate result
- That the sampling problems arising in industry and research are referred to specialists and not just to anyone.

Part I

THE QUALITATIVE APPROACH

ACHIEVING ACCURACY IN SAMPLING

Chapter 2

How Can the Mass of a Lot L be Reduced?

The sole aim of sampling is to reduce the mass of a lot L without significantly changing its other properties (this is the purpose of the quantitative approach). Reducing the mass of a lot of broken material can be done:

1. Either by taking increments: used on flowing streams of material (Section 2.1) or
2. By splitting: used when the whole of the lot can be handled (Sections 2.2 to 2.4).

2.1 INCREMENTAL SAMPLING

Incremental sampling is used on flowing streams of discrete particles (solid fragments, liquids, multi-phase mixtures) that can be characterised by one-dimensional time or space models. Sampling by increments is illustrated schematically in Figure 2.1; it will be studied in Chapter 5.

The 'increment I' is the quantity of material taken from the lot in a single pass of the sampling tool. Q is the number of increments $I_1 \ldots I_q, \ldots I_Q$ that are extracted from the lot L at intervals of time or distance that may or may not be regular.

That part of the sampling apparatus which is in contact with the material and which extracts the increment is called the 'cutter'.

The 'sample E' is the assemblage of all the Q increments I_q and the 'stock S' is what remains of the lot after the sample has been taken from it:

$$E = \sum_q I_q \text{ with } q = 1, 2, \ldots, Q$$

The 'mass sampling rate τ is defined as:

$$\text{Mass sampling rate} = \frac{\text{Mass of sample } E = \text{sum of } Q \text{ increments } I_q}{\text{Mass of lot } L}$$

In the same way, the 'time sampling rate τ' is the fraction

$$\text{Time sampling rate} = \frac{\text{Time taken to cut the } Q \text{ increments } I_q}{\text{Time taken by lot } L \text{ to pass the sampler}}$$

Incremental sampling generally has a low sampling rate (by mass or time) in the range of about 5% to 0.01%.

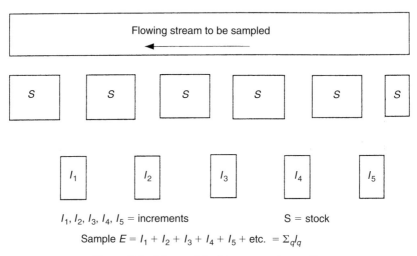

I_1, I_2, I_3, I_4, I_5 = increments S = stock

Sample $E = I_1 + I_2 + I_3 + I_4 + I_5 + \text{etc.} = \Sigma_q I_q$

Figure 2.1   The model of incremental sampling

2.2   SAMPLING BY SPLITTING

Splitting is particularly well suited to sampling a lot made up of discrete particles (solid fragments, molecules, ions) whose mass is small enough or whose intrinsic value is large enough to justify handling it either by hand or mechanically in its entirety at least twice. The sampling rate can be as high as 50%. Splitting divides the lot L into 2 or N twin-fractions that are regarded as having been taken under identical conditions. They are similar to each other except for small random differences.

Splitting into two twin-fractions using a large enough sample splitter or by alternate shovelling is shown in Figure 2.2; it gives a theoretical sampling rate of $1/2$. The actual rate is a random variable with a mean of $1/2$.

Splitting into N twin-fractions by true fractional shovelling is shown schematically in Figure 2.3. It gives a theoretical sampling rate of $1/N$, $2/N$, etc.

Splitting into two non-twin fractions by degenerate fractional shovelling is shown schematically in Figure 2.4. It gives a sampling rate of $1/N$.

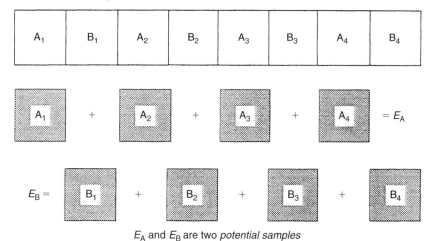

E_A and E_B are two *potential samples*

Figure 2.2   Splitting into two twin-fractions. An example of alternate shovelling

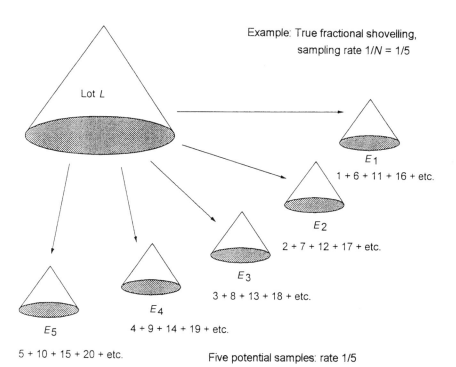

Figure 2.3   Sampling by true splitting into $N = 5$ twin-fractions

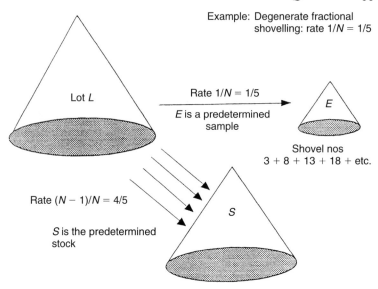

Figure 2.4 Deterministic sampling by degenerate fractional shovelling into two unequal fractions E and S; E[rate $= 1/N = 1/5$] and S [rate $= (N - 1)/N = 4/5$]

Important note: Splitting is not sampling, it is only a precursor to it. It begins with the division of the lot L into two or N fractions, twin or non-twin. It will be recalled that fractions are said to be 'twins' when they have been taken under similar conditions and differ by only a small random factor. Sampling proper begins only at the selection process, which enters at the second stage when some of the fractions are retained as samples. Thus, after splitting into two parts as shown in Figure 2.2, there are two non-identical pseudo-halves forming two potential samples E_A and E_B. It is the selection from these two, of which one is to be the actual sample, that lies at the heart of sampling. The method of selection has to be defined. It can be deterministic or random.

2.2.1 Splitting into Two Twin-fractions

Figure 2.2 is a schematic layout of splitting or alternate shovelling. Two potential samples E_A and E_B are obtained from the splitting process shown in Figure 2.2 Tossing a coin, for example, decides which of the two will be kept as the actual sample. This technique, which can be used in successive splittings, is of wide application in commercial sampling. The operator is often tempted to choose always the same pseudo-half. The introduction of a random element in the selection process, by coin tossing for example, is made necessary by the need to eliminate the two following risks (these two points are illustrated in Sections 22.7.3 and 23.3 of reference [7]):

- When splitting lots containing particles measuring more than a few milli-metres it is possible to do alternate shovelling in such a way as to put more of the coarser granules into one of the two fractions. When splitting is done repetitively, i.e. if the lot L is split n times into two fractions, the operator, although honest, has a natural tendency to choose the same fraction each time, and this increases by a factor of n the possible deviation between the 2^n fractions so obtained.

- A dishonest operator (and there are such) can choose which of the two splits seems to favour his agent more. For example, if he sees that fraction E_A contains more 'big' fragments and he knows that these are richer (or poorer) than the small ones, he can 'adjust' things if he wishes to.

2.2.2 True Splitting into N Twin-fractions

In Figure 2.3 shovelfuls are taken from the lot L and placed one by one, in order, onto the heaps $E_1 \ldots E_5$, and again $E_1 \ldots E_5$, and so on until the lot is exhausted.

As well as the manual methods there are mechanical rotating sample splitters that divide a lot L into N twin-fractions. The methods examined here allow the drawing of either one sample of size $1/N$ or one sample of any size between $2/N$ and N'/N (with $N' < N$), or several samples of size $1/N$ (if splitting is done with the object of making comparative studies).

The basic method, shown in Figure 2.3, illustrates the true fractional shovel-ling of a lot L with a sampling rate of $1/N$. Section 2.2.3 following shows how a method called degenerate fractional shovelling splits a lot L into two non-twin fractions, one with a sampling rate of $1/N$, the other with a rate $(N-1)/N$ whose properties are very different.

True fractional shovelling has this advantage: it is easy to make it an equitable sampling method by choosing by chance any one of the N potential samples that will be kept as the actual sample. It also has one disadvantage. It calls for a great deal of handling, and hence is costly.

2.2.3 Degenerate Splitting into Two Non-twin Fractions

Degenerate splitting is sometimes used to reduce costs, and is done by putting every Nth shovelful onto a heap that will be the final predetermined sample E, and the remainder onto a predetermined stockpile S of rejects. This operation is illustrated in Figure 2.4 for $1/N = 1/5$.

That sampling rates greater than $1/2$ are rarely used is accounted for by the fact that E is a predetermined sample and S is a predetermined stock. This suppresses the likelihood of a random selection and with it the guarantee that sampling is equitable. If the sampling operation is technically biased (and it is well known that the operator has a certain latitude in introducing a bias when sampling coarse materials in this way), it will inevitably be inequitable.

The method can be used with mechanical shovels to sample lots that would otherwise be considered unmanageable because of their high tonnages. The method was used by the author when he was called in to arbitrate on a 16 000 tonnes lot of ore at Pacific port. The rate $1/N$ was $1/20$ for primary sampling and $1/10$ for secondary sampling, using a five-tonne mechanical shovel. Third-stage sampling used a one-tonne shovel and a $1/10$ sampling rate. The final sample of eight tonnes was dealt with at the laboratory.

Sampling took place under the author's supervision, using people employed by neither the buyer nor the seller, and under conditions that made cheating impossible. The method is simple and because it is not excessively costly offers an acceptable solution, albeit one that is not used as much as it deserves to be.

Chapter 3

Heterogeneity and Sampling

Sampling errors arise first and last from the existence, in one form or another, of heterogeneity. It is therefore essential to analyse the concepts of homogeneity and heterogeneity in depth, qualitatively in this first part and then quantitatively in the following two parts.

As noted above, the sole object of sampling is to reduce the mass of the lot to be valued. Now, the lot is a set of constituent elements. This set can be a population or a series (time- or space-based). Mass reductions require selections to be made from sub-sets of constituent elements, as defined next.

3.1 CONSTITUENT ELEMENTS AND UNITS

The 'constituent elements' of a lot of material are the smallest elements that can be considered to be immutable in the physical, chemical and mechanical conditions of sampling. It will be assumed that the temperature remains constant during this operation. The constituent elements are:

- For solid particles: fragments
- For liquids and gases: molecules and ions. The theory can be applied to gases as well, but these pose practical problems that are not dealt with here.
- For multi-phase mixtures: fragments, molecules and ions.

A lot will be considered to be a set of 'units' and the study of heterogeneity will deal with the following 'units':

- Either separate constituent elements F_i (F = fragments) or
- Groups I_n (I for future increments) of neighbouring constituent elements.

3.2   CONCEPTS OF HOMOGENEITY AND HETEROGENEITY OF A SET

A set of units is:

- *Homogeneous*: when all its units are strictly identical to each other. 'Strictly' is emphasised. Approximation has no place in what follows; scientific rigour is indispensable and all definitions have to be adhered to with undeviating strictness.

Consequence No. 1: Sampling by selecting whole units from a set that is supposed to be homogeneous is therefore a structurally exact operation as the sample will consist of identical units, as do the units that make up the set. But . . .

  homogeneity is an abstract mathematical concept that does not exist in the real, material world.

- *Heterogeneous* when the units are not all strictly identical to each other. It might be argued that although this remark holds good for solid fragments, would they not be identical at the scale of molecules or ions?

A critical reader might claim that pure water is an example of homogeneous material. From its appearance on a large scale, yes! But, if it is looked at closely, no!. It is an indescribable, or nearly so, mixture of H_2O molecules, H^+, OH^-, O^{2-} ions, without mentioning all the possible isotopic combinations of hydrogen and oxygen, or ions of foreign material that are never totally absent.

Consequence No. 2: This leads to the conclusion, one that never fails to disconcert some readers, that:

  Heterogeneity is the only state in which a set of material units or groups of units can be observed in practice.

Consequence No. 3: Sampling a set of material elements by selecting whole units is therefore necessarily an operation that is structurally inexact and inevitably generates errors.

Consequence No. 4: It follows from consequences 1, 2 and 3 that every sampling error is the result, therefore, of the existence of heterogeneity in one form or another within the lot of material. It will be demonstrated that there are several forms of heterogeneity.

  Heterogeneity is seen as the sole source of all sampling errors

The difference between homogeneity and heterogeneity is not really qualitative, but is quantitative: homogeneity is the inaccessible condition of zero heterogeneity, so it is necessary to go more deeply into the concept of heterogeneity, as will be done in Chapter 8. Several types of unit and as many conditions of homogeneity/heterogeneity have to be defined:

- If the unit is a single constituent element: the concepts of constitutional homogeneity and heterogeneity apply.
- If the unit is a group of neighbouring constituent elements: the concepts of distributional homogeneity and heterogeneity apply.

Chapter 4

Sampling Definitions

4.1 INTRODUCTION

Everyone speaks of representative samples but no one defines them scientifically. For example, the following definition appeared in Section 2.2.1 of a work published in 1995 [13]:

> Representative sample: This is a sample that is typical of the parent material for the characteristic under inspection.

A sample is representative if it is typical! There are many examples in this genre, and the one given here is only the most recent of them!

To make an objective, scientific definition of a representative sample, it is first necessary to define the properties of 'selection', 'sampling' and of 'sample taken' as functions of:

- Either the conditions prevailing before the sample was taken, and which can be controlled to some extent. This is discussed in Section 4.3 or
- The results of sampling, i.e. the statistical properties of the error observed after taking the sample. This is the object of Section 4.4.

4.2 THE ROLE OF THE THEORETICIAN

Designers, manufacturers, users of sampling equipment, have no means of directly influencing the results, i.e. the errors, of sampling. They do not know how to design a sampling machine or a sampling method that produces a sample which is 'accurate', 'reproducible' or 'representative'. It is in vain to write, as one sees so often particularly in the standards, that 'the analyst must work on representative samples only', because he has no direct means of doing so. He only knows the conditions under which the sample was prepared. It is therefore these conditions that must be defined first of all. The role of the theoretician is thus:

- To build a bridge, on firm theoretical foundations, linking the conditions of the selection process to the resulting sampling errors
- To explain to the user what he can expect or ask from the manufacturer
- To explain to the manufacturer what he must do—or avoid doing—if he wants his products to take representative samples.

4.3 DEFINITIONS OF SELECTION; OF SAMPLING; AND OF A SAMPLE AS FUNCTIONS OF THE CONDITIONS UNDER WHICH SELECTION TOOK PLACE

The following definitions were first proposed in 1975 and have been accepted at least by those aware of them. A selection is called:

- *Non-probabilistic*: When certain constituent elements of the lot to be evaluated have a zero probability of being taken into the sample.
- *Probabilistic*: When all the constituent elements of the lot to be valued have a non-zero probability of being taken into the sample.

A probabilistic selection can be correct or incorrect. It is:

- *Correct*
  – When all the constituent elements of the lot have an equal probability of being taken into the sample
  – When the increments and the sample are not altered in any way.
- *Incorrect*: When at least one of these two conditions is not fulfilled. The probability of selection is then affected by the physical properties of the constituent elements (size, density, shape etc.).
- *Non-correct*: When selection is either non-probabilistic or probabilistic but incorrect

The difference between 'samples and specimens': anticipating a later chapter, henceforth samples and specimens will be defined thus:

- Only a correct selection produces samples that are structurally accurate.
- A non-correct selection produces specimens that are structurally biased.

4.4 HOW TO MAKE A CORRECT SELECTION

This section is of prime importance to those wishing to understand why sampling is not a simple process of material collection. It was noted above that a selection is said to be correct when all the constituent elements of the lot have an equal probability of being taken into the sample. There are two ways of making a correct selection.

4.4.1 No Distributional Hypothesis is Made

Whatever the distribution of the constituent elements between the elements of volume in the domain occupied by the lot which may be selected as increments, selection is correct if and only if all these elements of volume have an equal probability of being selected.

4.4.2 The Lot has been Homogenised (which does not imply that it is 'homogeneous')

It will be shown in Chapter 8 that it is mathematically impossible to eliminate the distributional heterogeneity caused by the spatial distribution of the constituent elements, but it can be reduced and—at least on paper—minimised by homogenisation. The distributional heterogeneity is minimal (Section 4.4.7 of reference [7]) when any correlations between the positions of the constituent elements and their physical properties (density, size, shape) have been eliminated. Where they do exist, such correlations are most often due to the differential action of gravity. When the correlations have been destroyed by homogenising, all the constituent elements are said to be distributed at random throughout the body of the lot.

Homogenising tends to give each constituent element an equal probability of falling inside each element of volume within the body of the lot.

Correctness is assured whatever elements of volume are selected, even, for example, only the most accessible ones.

However, it would be foolish to assume that mixing has produced an unrealisable distributional homogeneity because, as shown in Section 3.2, the implicit introduction of such a hypothesis would allow all sampling problems to be resolved by assuming they have disappeared. This is why the words 'homogeneous' and 'homogeneity' should be banned from sampling as they have no foundation in reality. It is possible to speak of a substance being 'homogenised', knowing that its heterogeneity has only been reduced, maybe even minimised, but never that it has been eliminated. Readers should distance themselves from those who postulate the essentially unattainable condition of homogeneity; such a hypothesis embraces all the methods that are based on the doubtful principle of 'catch what you can, provided it doesn't cost too much'.

4.5 DEFINITIONS OF SELECTION; OF SAMPLING; AND OF A SAMPLE AS FUNCTIONS OF THE SAMPLING ERROR

4.5.1 Nomenclature

a_L the actual but unknown proportion by mass (i.e. the decimal portion) of component A in the lot L

a_E the actual but unknown proportion by mass of the component A in the sample E

e the relative sampling error, defined as:

$$e = (a_E - a_L)/a_L$$

The error e is a random variable which as such can be characterised by its distributional law, usually normal (maybe not so for trace elements) and by its moments.

4.5.2 Definitions

See also Appendix 1. A selection is or should be (the conditional mood applies to the first two of the following definitions):

- *Exact*: If error e is strictly identically zero, implying $a_E = a_L$: this identity is never observed structurally
- *Strictly accurate*: If in this case (property of the mean): $m(e) = 0$: this identity is never observed structurally
- *Accurate* (in practice): In this case (property of the mean) $|m(e)| <= m_0$ where m_0 is the maximum bias allowed
- *Biased*: When this condition is not satisfied (property of the mean) $|m(e)| > m_0$: the bias exceeds its maximum allowed value
- *Reproducible*: In this case (property of the variance) $s^2(e) <= s_0^2$: s_0^2 is the maximum allowed variance
- *Representative*: In this case (property of the mean square, a composite quantity involving both the mean and the variance) and defined as $r^2(e) <= r_0^2$ in which, by definition:

$$r_0^2 = m_0^2 + s_0^2$$

In other words a sample is representative when it is taken by a selection method that is both accurate and reproducible.

Thus representativeness is characterised by the absence of bias and an acceptable variance. As far as the author is aware, this is the only possible objective and scientific definition of representativeness.

4.6 RELATIONSHIPS BETWEEN SAMPLING CONDITIONS AND THE STATISTICAL PROPERTIES OF THE SAMPLING ERROR

To justify these definitions we must anticipate the results of the quantitative theory developed in Part II and the results are summarised as follows:

1. A correct sampling method is always structurally accurate. In addition, its variance is minimal so that its representativeness is maximal.
2. The correctness of a sampler is the result of its design, construction, installation, usage and maintenance, as well as the absence of any changes in the increments it takes and of any changes in the final sample, which is a property of the sampling equipment, of sample preparation, transfers, comminution, etc. These conditions are simple and easy to satisfy and are described in Section 6.2.7.
3. Non-correct sampling is always structurally biased. It may be accurate over short periods, but these cannot be forecast and so are unusable. This makes the tests of accuracy recommended by certain standards (the so-called bias tests) not only useless but also dangerous as they offer a false sense of security. What was observed on day D at time t cannot be extended to day D' and time t'. Furthermore, the variance is no longer minimal; additive variances sum to the variance observed when sampling is correct. All non-correct samplers are implicitly based on the unrealistic hypothesis of distributional homogeneity.
4. To be representative a sample must first be correct, and this implies that it is accurate. But this necessary condition is not by itself sufficient. As well as having a negligible bias, representativeness requires reproducibiltiy, i.e. a minimum variance, which itself depends on the quantitative properties of the sample (e.g. the mass and number of increments). Minimising the variance will be studied in Parts II and III.

4.7 BIASES INTRODUCED BY INCORRECT SAMPLING

From the author's years of experience and from hundreds of inspections of industrial plants and laboratories, the following biases are given as fairly typical values:

Primary sample (non-probabilistic): up to 1000%
Secondary sampling (probabilistic but incorrect): up to 50% (and probably much more)
Analysis: 0.1–1.0%

Thus it is pointless and illusory to return an analytical result to three or four supposedly significant decimal places if the sample analysed is insufficiently representative and even more pointless if it is biased.

4.8 ACCURACY AND REPRODUCIBILITY: THE DIFFERENCE

In about 1960 the maker of a certain sampler and a potential client delivered seven samples taken under the conditions illustrated in Figure 4.1. The sampling rate was of the order of 1%. The seven samples were taken consecutively (but not under the author's supervision) from a lot L of some 500 kg of iron ore. The first stage resulted in a sample E_1 and stock S_1. S_1 became the lot L_2, which was fed to the same sampler to give a second sample E_2 and a second stock S_2, and so on to sample E_7 and stock S_7. Samples E_1 to E_7 were sent to the author to prepare and analyse. He also asked that the stock S_7, which had been put to one side, should be analysed as well. As a result of this last analysis it was possible to calculate Table 4.1 using the expression $S_6 = L_7 = 0.99S_7 + 0.01E_7$ (from the sampling rate of 1%). The analytical errors were considered negligible in comparison to the sampling bias.

Mean of the sample grades: 51.476%Fe
Standard deviation: 0.037%Fe
Relative standard deviation: 0.07%

First conclusion: There seems to be good reproducibility.
Mean bias: +0.597%Fe (or 1.17% relative).
Second conclusion: Accuracy is far from being assured, so the samples, being biased, are far from being representative.

The sampler was banned. Confusing reproducibility with accuracy, the client had been prepared to install the sampler. The last stock S_7 was analysed in order

Table 4.1

Lot no.	Calculated grade (Fe%)	Sample E	Analytical grade (Fe%)	Observed bias $(E - L)$ (Fe%)
$L_1 = L$	50.897	E_1	51.450	$E_1 - L_1 = +0.553$
$L_2 = S_1$	50.886	E_2	51.435	$E_2 - L_2 = +0.544$
$L_3 = S_2$	50.886	E_3	51.485	$E_3 - L_3 = +0.599$
$L_4 = S_3$	50.880	E_4	51.470	$E_4 - L_4 = +0.590$
$L_5 = S_4$	50.874	E_5	51.485	$E_5 - L_5 = +0.611$
$L_6 = S_5$	50.868	E_6	51.550	$E_6 - L_6 = +0.682$
$L_7 = S_6$	50.861	E_7	51.460	$E_7 - L_7 = +0.599$
S_7			50.855	

to demonstrate the bias. With forty years of hindsight the author now believes that the seven samples sent for analysis were not the actual samples collected by the sampler but were instead substitutes. Such biases are, in fact, always accompanied by high standard deviations, and never in the author's experience has he encountered samplers introducing practically constant biases. However, thanks to an effective publicity campaign this sampler was used widely at iron mines and cement factories.

Figure 4.1 shows the sampling experiment described previously and the unfortunately all too frequent confusion between accuracy and reproducibility.

4.9   NON-PROBABILISTIC SAMPLING: 'PICKING'

This is also called 'grab sampling'. It is simple, even simplistic, its principle being:

The most accessible fraction of the lot is taken because it is the cheapest in the short term.

All picking methods give non-probabilistic selections. There are sampling appliances that use them, implicitly, without their designers, manufacturers or users being aware of the fact that all they have are simple material-collecting devices.

A non-probabilistic sampling theory is an impossibility because part of the lot is inaccessible to the sampling tool. The errors it generates are considerable and may be ruinous in the long term. Non-probabilistic sampling must always be avoided; nevertheless, it is very widespread.

4.9.1   Sampling by Picking from the Tops of Container Loads (Wagons, Trucks, Casks, Drums, etc.)

Some shovelfuls (the increments) are taken from the surface because it is a cheap way to take a 'sample'. Its users appear unaware that the method is based on the hypothesis that the vertical distribution of the constituent elements is homogeneous. This hypothesis is dangerous because, as has been noted, it is unrealistic.

A sample is made up of the increments (Figure 4.2). With liquids or multi-phase mixtures, the shovel is replaced by a bucket or similar. In the case of solutions of heavy metals, or with multi-phase mixtures, segregation under the differential effects of gravity is the rule, not the exception, and serious biases may follow.

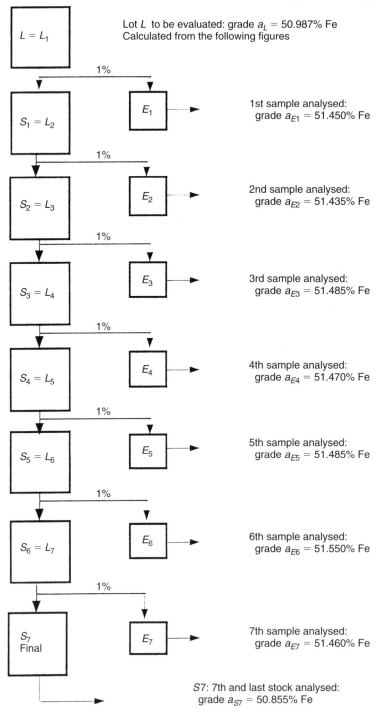

$L = L_1$

Lot L to be evaluated: grade $a_L = 50.987\%$ Fe
Calculated from the following figures

1%

$S_1 = L_2$

E_1

1st sample analysed:
    grade $a_{E1} = 51.450\%$ Fe

1%

$S_2 = L_3$

E_2

2nd sample analysed:
    grade $a_{E2} = 51.435\%$ Fe

1%

$S_3 = L_4$

E_3

3rd sample analysed:
    grade $a_{E3} = 51.485\%$ Fe

1%

$S_4 = L_5$

E_4

4th sample analysed:
    grade $a_{E4} = 51.470\%$ Fe

1%

$S_5 = L_6$

E_5

5th sample analysed:
    grade $a_{E5} = 51.485\%$ Fe

1%

$S_6 = L_7$

E_6

6th sample analysed:
    grade $a_{E6} = 51.550\%$ Fe

1%

S_7
Final

E_7

7th sample analysed:
    grade $a_{E7} = 51.460\%$ Fe

$S7$: 7th and last stock analysed:
    grade $a_{S7} = 50.855\%$ Fe

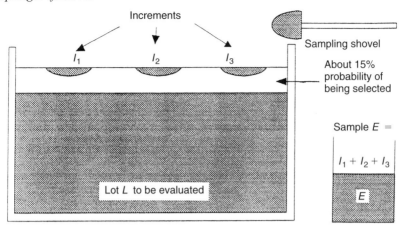

Figure 4.2 General principle of 'picking'. Increments taken from the top of a wagon-load

4.9.2 Sampling by Picking from the Side or Top of a Stream of Material on a Belt Conveyor

It was this kind of sampling (Figure 4.3) that caused the bias of 9% mentioned in Section 1.7.3.

4.9.3 Sampling by Picking from a Layer of Powder

This is one of the methods used by analysts when taking the final analysis pulp from a sample sent to them (Figure 4.4). It is not necessarily the worst provided the powder has been well mixed beforehand. In general it is to be preferred to the automatic samplers suggested by some manufacturers of analytical apparatus and balances.

4.10 THE DANGEROUS ASSUMPTION OF HOMOGENEOUS DISTRIBUTION: HETEROGENEITY FROM FRAUD OR BY GRAVITATIONAL SEGREGATION

4.10.1 Sampling Truckloads of Sugarbeets

The author was asked by the director of a sugar refinery to determine the cause of a serious difference between the tonnage of rocks recovered from the plant's

Figure 4.1 Experiment illustrating the confusion between accuracy and reproducibility

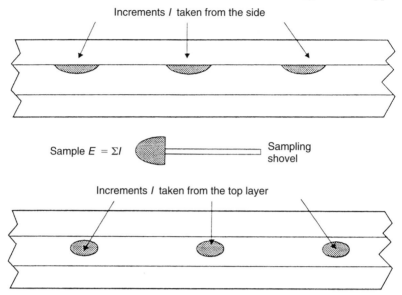

Figure 4.3   Picking from a belt conveyor

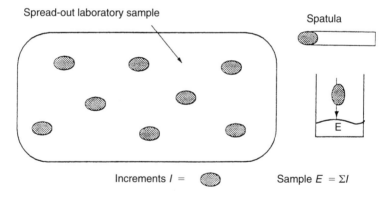

Figure 4.4   Picking from the surface of an evenly spread layer of powder

waste bins during the refining process and the tonnage of rocks determined by sampling each truckload as it entered the refinery. The sampler was mechanically very sophisticated (which inspired some confidence). It was operated hydrauli- cally by forcing three vertical probes fitted with sharp cutters of square section down through the beets to the bottom of the trucks. The probes were sited at random over each truck with the aid of a computer. The open cutters passed through the beets on their way to the bottom of the truck, and were then shut.

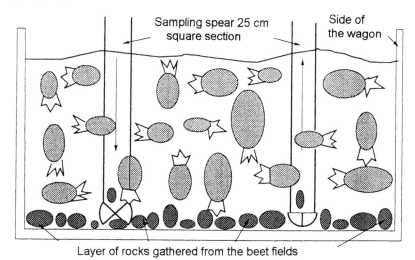

Sampling spear 25 cm square section

Side of the wagon

Layer of rocks gathered from the beet fields

Figure 4.5 Sampling wagonloads of sugarbeet by sampling spears

Figure 4.5 represents a vertical section through a truck showing, on the left, the open cutter passing down through the beets and on the right, the closed cutter withdrawing the increment. Horizontally, there is an equal probability of selection. Correct sampling requires that all of the vertical section is recovered in its entirety, but the cutters did not reach to the bottoms of the trucks. It follows that sampling is incorrect and virtually non-probabilistic because the elements of the bottom-most layer have only a small probability of being taken. Taking advantage of this design fault the growers saw that instead of having to stack the rocks around the edges of their fields they could instead lay them on the truck bottoms, thus avoiding a reduction in the area available for cultivation and any further damage to their ploughs. Shrewd but deceitful!

4.10.2 Samples Delivered by Feeder Tube to an AA Analyser

The manager of a certain gold mine called in the author to find out why there was a serious difference between the weight of gold calculated from the analyses of pregnant gold solutions and the weight of gold actually recovered (Figure 4.6). Having checked the primary sampling procedure the author was taken to the laboratory where the sample had been left on a bench some eleven hours previously. Having carefully calibrated the AA machine, the operator inserted the feeder microtube into the flask. First reading, 387, which was converted into g/l. The flask was then swirled gently in slow circles. Second reading, 550. A clean cork was inserted and the flask shaken vigorously. Third reading, 850. Several shakings and readings followed that stabilised at 850 ± 3. The minimum heterogeneity had thus been reached.

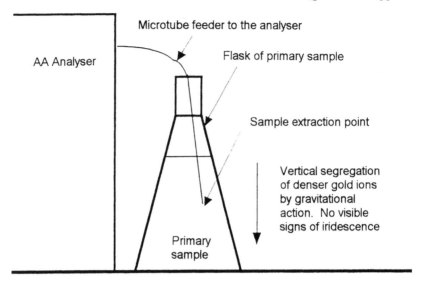

Figure 4.6   Laboratory sampling: feed to an AA analyser

Explanation: the solution had remained at rest for several hours and the gold cyanide ions, being denser than the water molecules, had segregated progressively to the bottom of the flask without leaving any visible trace of having done so.

Conclusions
1. The fact that a solution seems clear (absence of solids) and limpid (no iridescence) does not prove that its ions and molecules are distributed homogeneously throughout. Invisible as they are, the latter segregate under differential gravitational action just as solid particles do.
2. Although trained as a graduate chemist, the operator had never heard of sampling and because of this was unaware that here was a situation in which something could go amiss. Furthermore, the instruction manual that came with the machine made no mention of the need to ensure the sample was thoroughly mixed. Double fault!

Chapter 5

Objects in Three, Two, One or Zero Dimension(s)

5.1 OBJECTS IN THREE DIMENSIONS

Strictly speaking, all objects occupy a Cartesian space of three dimensions. Indeed, one particular problem is that of sampling a three-dimensional object that cannot be reduced to a smaller number of dimensions. Among these are to be found compact objects of large dimensions such as orebodies, for the evaluation of which a theory has been developed by Georges Matheron [14] and his disciples, e.g. Michel David [15].

In the same category lie the static stockpiles found at pitheads; on quays for loading and unloading ore shipments, etc. They are too heavy to be reduced to the form of a zero- or one-dimensional object that can be sampled easily and correctly, for example on a conveyor belt or by a mechanical shovel. Also in this category are the contents of grain silos; the disposal of rubbish and other public waste; industrial waste that has to be sampled to check on its possible environmental impact, etc.; and even the sampling of oceans, seas and lakes.

Matheron created the science of geostatistics to solve some of these problems, and it is now taught on a worldwide scale and spread by numerous congresses and publications. For this reason the problems addressed by geostatistics will not be dealt with here, and the reader is referred to its literature, which is plentiful.

5.2 OBJECTS IN TWO DIMENSIONS

For purely practical reasons three-dimensional objects are in fact sampled as though they were two-dimensional objects on a horizontal plane, from which 'drill cores' are taken and analysed section by section. This category includes stockpiles of solid fragments whose thicknesses are more or less uniform; forestry, and so on.

5.3   OBJECTS IN ONE DIMENSION

These are linear objects, linear in space or time, such as those that are studied in Chapter 6 and quantitatively in Chapters 11 and 12. Spatially, it can be applied to natural objects such as rivers or to manufactured objects such as iron castings, copper wire bars and the like. Temporally, the object is a flowing stream of material, such as the solid fragments, liquids, multi-phase mixtures, found in many continuous industrial operations and port-handling facilities.

When one of these has to be sampled by increments, it must be treated as a chronological or geometric series of units. The order of the units is of paramount importance and it must be taken into consideration because of the possibility that some degree of autocorrelation may exist between them. To treat an ordered series as if it were a non-ordered population is a serious mathematical error, an error blithely committed by many statisticians and certain standards committees (with the notable exception of those of ISO-TC 183). The reader will find more useful information on this topic in Appendix 2.

The quantitative theory for the sampling of one-dimensional sets that are assumed to be ordered is presented in Part III.

5.4   OBJECTS OF ZERO DIMENSION

A convenient convention is to describe a population as a set of units of the same nature, that have been put together regardless of any order that may or may not be present. This is not an arbitrary convention; a one-dimensional object can be likened to a vector, a zero-dimensional object to a scalar. It will be seen that whereas the latter can be characterised by a variance, the former needs to be characterised by a function called the variogram.

The quantitative theory for sampling populations of units that are not presumed to be ordered is presented in Part II.

5.5   MEETING POINT OF THE ZERO- AND ONE-DIMENSIONAL MODELS

What are called 'models of heterogeneity and of sampling in zero and one dimension' will be developed in the following chapters. The two models neither oppose nor contradict each other. They are complementary.

Zero-dimensional models characterise the small-scale properties of all the objects, as if they were viewed through a magnifying glass. At the scale of the constituent elements, large-scale correlations do not intervene and they can be regarded as being negligible, even if they are not neglected in the models used to characterise them.

The one-dimensional models of heterogeneity and sampling characterise the large-scale properties of all the objects, mainly those in flowing streams of material, as if they were viewed through a wide-angle lens. Correlations intervene and must be taken into consideration.

Chapter 6

The Practical Sampling of Moving Streams of Material

This chapter deals with materials such as solid fragments, liquids and multi-phase mixtures in motion along conveyors, pipes, chutes, etc., so as to form a one-dimensional object. The problem is of great practical importance. The different options will be examined and assessed on the basis of the criteria set out in Chapter 4.

Stream and river sampling poses more particular problems that will not be dealt with here. When it is a question of controlling their pollution it is the source of such that needs to be sampled, in the knowledge that nothing can be done about the occasional deliberate discharge of pollutants.

6.1   THREE WAYS OF REDUCING THE MASS OF A MOVING STREAM

These methods are illustrated in Figure 6.1.

6.2   SAMPLING ALL THE STREAM FOR A FRACTION OF THE TIME

6.2.1   Taking Increments from a Stopped Belt (Figure 6.2)

Stopped-belt sampling applies only to solid fragments, usually more or less dry, but sometimes quite wet. It is the nearest thing in practice to the point increment model that will be studied in Section 6.5. Increments are taken in the sequence shown in Figure 6.6.

The frame is made of two sheets of stout sheet metal that follow the contour of the belt and are kept parallel to each other by two bracing bars. To take an increment the belt is stopped and, opposite a fixed marker point, the frame is

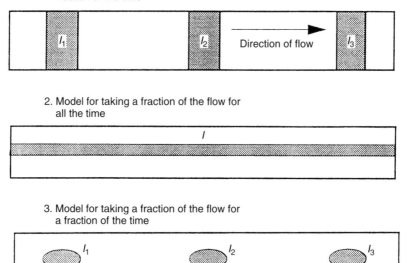

Figure 6.1 Three ways of reducing the mass of a moving stream of material

forced into the material until it bottoms onto the belt. The increment is the material between the cheek plates. However, the author has never visited any industrial installation using conveyors, often very long ones, where he felt able to recommend the use of stopped-belt sampling and its attendant periodic stopping of the belt. In some port installations belts are several kilometres long and operate at high capacities (world record in 1996: 16 000 t/h with peaks of

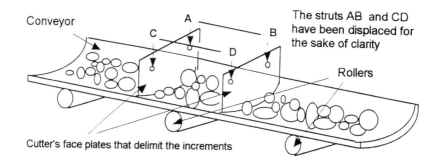

Figure 6.2 Stopped-belt sampling

20 000 t/h). The problem is not that of stopping the belt but of starting it again without risking burning out an expensive drive motor of several hundred kilo-watts. Nevertheless, this type of sampling is recommended by several standards committees as a 'reference sampling method' to test the accuracy of industrial samplers (bias tests).

Some people of a practical mind are aware of the problem and have designed transverse samplers, so called because they cut across the stream at a transfer point. There are a number of them and they can be put into two main groups: straight-line cutters that sample in' both directions (Figure 6.3) and rotating cutters that cut in the same direction (Figure 6.4). Many variations of both types have been described in the author's previous publications. In practice, the cutter discharges into some transport system that can be another moving belt or a vibratory feeder (for more or less dry solids), or a system of launders and pipes (for liquids or multi-phase mixtures).

6.2.2 Straight-path Transverse Sampler: Bi-directional

The sample cutter moves horizontally between two rest positions (Figure 6.3(2)) fixed by stops. A reduction motor moves the cutter alternately from left to right and then from right to left. The machines are usually fitted with timers to adjust the intervals between successive cuts. There are a number of variations; some cutters move from front to rear instead of from right to left. They provide an alternative cut that, theoretically, is equivalent to the one described in Figure 6.6(4).

6.2.3 Uni-directional Transverse Linear Sample Cutters

Other variations are designed to cut the stream by moving always in the same direction. The cut so produced is shown schematically in Figure 6.6(5).

6.2.4 Uni-directional Transverse Rotary Sampler

There are innumerable variations on this design. The classic example is illustrated in Figure 6.4, and the cut it produces is schematised in Figure 6.6(5).

6.2.5 The Mathematical Integration Model

The lot L of actual but unknown grade a_L that is to be estimated passes between times $t = 0$ and $t = T_L$. The flow rate is assumed to be constant. Let $a(t)$ be the grade of the material that has passed between times t and $t + dt$. By definition:

$$\text{Area } S = \text{ABCD} = \int_0^{T_L} a(t).dt = a_L T_L \text{ whence } a_L = \frac{1}{T_L} \int_0^{T_L} a(t).dt$$

1. Vertical section

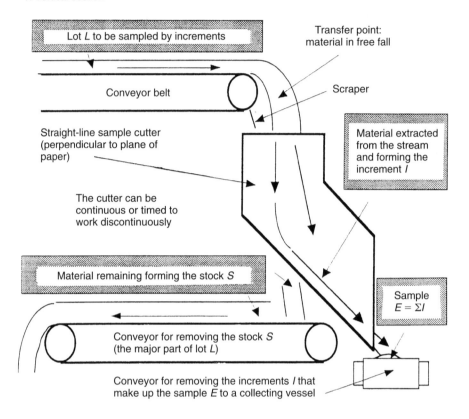

2. Horizontal section: bi-directional cuts made at speed *V*

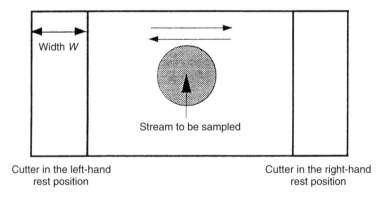

Figure 6.3 Straight-path transverse bi-directional sampler

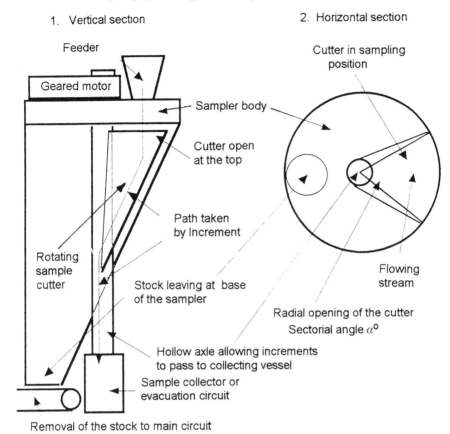

Figure 6.4 Uni-directional transverse rotary sampler

All transverse samplers depend on a mathematical model called 'integration by points', as illustrated in Figure 6.5. It is an approximate method that replaces the area $S = $ ABCD by $S' = $ A′B′C′D′.

6.2.6 Mathematical Model of the Actual Cutting Process

In Part III, a mathematical model is developed to estimate the moments of the relative integration error of the random variable IE:

$$IE = (a_E - a_L)/a_L = (\text{Area A′B′C′D′} - \text{Area ABCD})/\text{Area ABCD}$$

It uses the true, but always unknown, value $a(t_q)$ of each theoretical point increment I_q comprising the elementary strip of material flowing past a fixed reference (the area swept by the cutter gape) between times t_q and $t_q + dt$ with $q = 1, 2, \ldots, Q$. To move on from theoretical point increments to real samples

1. Actual but unknown curve of the grade function $a(t)$

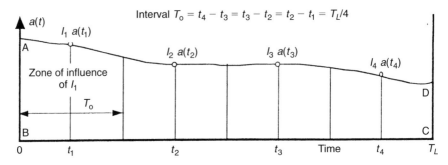

2. Stepped line as an approximation of the grade function $a(t)$
   The area $S' = A'B'C'D'$ is an estimator of the area $S = A\,B\,C\,D$. Each
   zone of influence is replaced by a rectangle R_1, R_2, R_3, R_4

$$S' = a(t_1)\,T_o + a(t_2)\,T_o + a(t_3)\,T_o + a(t_4)\,T_o = 4\,a_E\,T_o = a_E T_L$$

$$a_E = \frac{[a(t_1) + a(t_2) + a(t_3) + a(t_4)]}{4} = \frac{1}{Q}\,\Sigma\,a(t_q)$$

a_E = the grade of the sample E and Q is the number of increments

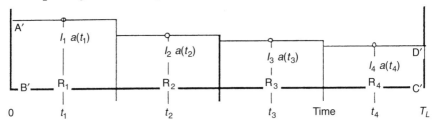

Figure 6.5 Mathematical model of point integration

made up of groups of neighbouring constituent elements requires several logical steps:

- *First stage.* Passing from the point increment I_q to the extended increment I'_q of duration Δt between times $t_q - \Delta t/2$ and $t_q + \Delta t/2$, i.e. from stage 1 to stage 2 of Figure 6.6.
- *Second stage.* From the extended increment I'_q in one dimension to the static increment I''_q in three dimensions, i.e. from stage 2 to stage 3 in Figure 6.6.

The latter is the stopped-belt sampling model. Thus, the ideal point model has

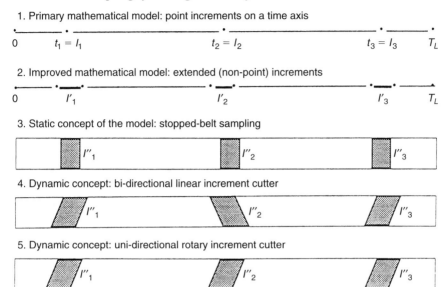

1. Primary mathematical model: point increments on a time axis

2. Improved mathematical model: extended (non-point) increments

3. Static concept of the model: stopped-belt sampling

4. Dynamic concept: bi-directional linear increment cutter

5. Dynamic concept: uni-directional rotary increment cutter

Figure 6.6 Integration model: materialisation of point increments

been developed so as to arrive at the reality of samples having real volumes comprising groups of fragments.

6.2.7 Correct Conditions for Transverse Sample Cutters

To be certain that an increment has been taken correctly, the cutter must collect each strand of the stream for the same length of time, although not necessarily at the same instant. To ensure this, three groups of conditions must be respected.

1. Conditions for correct delimitation (Chapter 10 of reference [7]):
   - The cutter speed V: constant during the whole traverse
   - The cutter width W: the sides of a straight-line cutter must always be parallel; those of a rotating cutter, radial.
2. Conditions for correct extraction (Chapter 11 of reference [7]):
   - The uniform cutter speed V must not exceed 0.6 m/s
   - The width W of the sample cutter must:
     - For solid fragments (> 3 mm) be at least three times the diameter d of the largest fragment: $W > 3d$
     - For fine solids, liquids, and mixtures of them: $W > 10$ mm.
3. Conditions for correct preparation (Chapter 12 of reference [7]): The integrity of increments and samples must be respected. Note: With high flow rates safety factors must be used.

(a)   *Respecting conditions 1 and 2 for correct delimitation (see Figure 6.7)*

When the two conditions of correct delimitation are respected, and only then, the static cut represented in Figure 6.6(3) by the increments I''_1, I''_2, I''_3 (stopped-belt sampling) is a rectangle, and the dynamic cuts in Figures 6.6(4) and (5) are parallelograms. The change from a rectangle to a parallelogram is because the strands of the moving stream are cut successively and not at the same instant. This is simply an approximation that follows from the practical implementation of the mathematical model and is of no consequence.

If one of these rules is not respected, for example if the speed V is not uniform, the cut CDEF is no longer a parallelogram and it can take any form whatsoever, for example C'D'E'F' or C''D''E''F''. A delimitation error or cutting error DE occurs (Figure 6.7).

C'D'E'F' results when the cutter is driven by an underpowered motor and slows down under the weight of the material entering it (differential oversampling). C''D''E''F'' is the result of the cutter still accelerating as it enters the stream, not having yet attained its proper speed V.

When the correct delimitation conditions are satisfied the delimitation error DE is identically zero.

• Primary mathematical model: point increments I (one dimension)

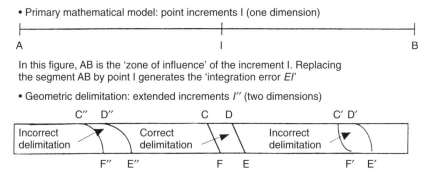

In this figure, AB is the 'zone of influence' of the increment I. Replacing the segment AB by point I generates the 'integration error EI'

• Geometric delimitation: extended increments I'' (two dimensions)

Figure 6.7 Point increment I and delimitation of the actual increment I''

(b)   *Respecting conditions 1 and 2 for correct extraction (Figure 6.8)*

In the following, it will be assumed that the geometric delimitation of the increment is correct. However, whether done correctly or not, delimitation defines only the volumes of the increments. It now remains to define discrete increments and to establish a logical relationship between these and extended increments. It is the rule of the centre of gravity, or the rebound rule.

All the fragments whose centres of gravity fall within the parallelogram CDEF belong rightly to the discrete increment I''. Fragments whose centres of gravity

• Correct delimitation: extended increment *I'* (two dimensions)

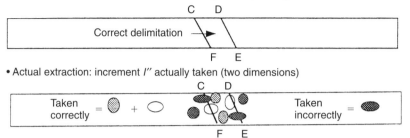

• Actual extraction: increment *I''* actually taken (two dimensions)

Figure 6.8 Actual extraction when delimitation is correct

fall outside CDEF belong to the stock. The centre of gravity rule arises because a fragment falling vertically and without spin onto a horizontal edge bounces to the side that contains its centre of gravity. Thus, each fragment can be considered as if it had been condensed into a point at its centre of gravity. When this simple rule is violated extraction is incorrect and a new error is created: the extraction error EE. The main extraction error occurs when a fragment after bouncing off the edge DE lands on the other edge CF and finally comes to rest in the stock to the left of CF. This occurs if the speed *V* of the cutter is too high (failure to respect rule 1 for correct extraction), or if the cutter width *W* is too narrow (failure to respect rule 2), or both.

When the conditions of correct extraction are all satisfied the extraction error EE is identically zero.

(c) *Conditions to be respected for correct preparation*

It will be assumed in what follows that delimitation and extraction are correct. The word 'preparation' should be thought of as a blanket term covering all the operations to which the individual increments and the sample will be subjected after their extraction from the lot, other than the interactions between the sample cutter and the moving stream (delimitation) and between the sample cutter and the constituent elements (extraction). The 'preparations' cover a large number of different operations: transfers between points; crushing and grinding of solid fragments; homogenising; drying; anything done by the sample preparers and analysts. All these operations are capable of generating errors that will be referred to as 'preparation errors PE'. Salting and all forms of tampering with the samples can be regarded as preparation errors.

When the integrity of samples and increments is strictly respected, the preparation error PE is identically zero.

Preparation errors PE come from six main sources:

1. The entry of foreign material or contamination. Examples: from dust; from foreign material present in the appliance used for taking or preparing the sample; from abrasion of the crushing and grinding machines; from corrosion of the equipment (rust); etc. (see Figure 6.9 and Section 12.3 of reference [7]).

2. The removal or loss of material. Examples: loss of dust during transfer; fragments bouncing out of the circuit; sample residues left in a machine because of incomplete cleaning; particles left wedged in the meshes of a sieve; etc. (see Figure 6.9 and Section 12.4 of reference [7]).

3. Alterations in chemical composition. In paragraphs 1 and 2 above it was noted how the addition or removal of any of the constituent elements resulted in failures to observe the rules for correctness. The same things done to the atoms and molecules of the elements themselves have the same effect.

   The grade, i.e. the proportion by weight of the element of interest, is the mass of that element divided by the mass of the active components, that is, the components that enter into grade calculations. For example, the grades of solids are, with rare exceptions such as phosphates, reported on the dry basis and not as they occur in their raw state, and analyses are accordingly run on dried material.

   The grade is altered if either term of the quotient is changed. Examples: oxidation of sulphides—the addition of oxygen to the numerator and denominator in different proportions; absorption of water or carbon dioxide by some oxides and hygroscopic chemicals; loss of water of hydration; loss of CO_2 by overheating during the drying of some carbonates; etc. Drying can be a highly sensitive operation.

4. Alterations in physical composition. Humidity and size analyses fall into this

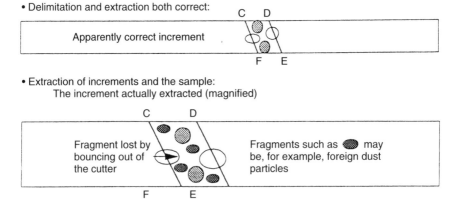

Figure 6.9 The resulting increment when delimitation and extraction are both correct. Examples or errors of the kind defined in (a) and (b) above

group. Examples: the addition of water by exposure to the atmosphere in equatorial climates, to rain and to sprays; creation of fines by rough handling—a potentially serious error. Loss of water by exposure to a source of heat such as the sun in tropical climates, or by being left too close to an industrial source such as a kiln. In the particular case of sulphur, which sublimes at 80 °C, drying under the usual conditions of 105–110 °C causes a loss of sulphur and hence an error of this kind.

5. Involuntary faults committed by the operator. The accent is on the adjective 'involuntary'. All too often sampling operators, although acting in good faith, lack even the most elementary qualifications and hence are likely to make errors because of ignorance, clumsiness or negligence; e.g. mixing fractions belonging to different samples, mis-labelling bottles, dropping samples and only partially recovering them, etc.: the list is far from exhausted. A recognised training course with the title of 'higher-grade sampling technician' is suggested.

6. Deliberate faults committed by the operator. The accent is now placed on 'deliberate'. Sampling is the weakest link in the chain leading to the determination of quality. It is also the least understood and therefore the one most vulnerable to fraud. Apart from the 'salting' of gold ores, the two areas in which the author has seen fraud attempted are in commercial sampling and in the control of possibly injurious effluents, particularly from industries handling uranium and other radioactive materials.

Note: The reader may argue that these operations do not belong to sampling proper, and to some extent this is true. However, it is no less true that they are undertaken in the context of sampling, either at the industrial level or in the laboratory. They are perpetrated daily by sampling and laboratory personnel and the errors they introduce affect the results to the same extent as true sampling errors, and their inclusion is justified.

6.2.8   Observations on the Correctness of Transverse Samplers

Sampling theory (see references [6] and [7]) shows that at each stage of sampling the total sampling error TE is the sum of the integration error IE and the materialisation errors ME that result from the transformation of the point increments of the mathematical model into material increments, thus:

$$TE = IE + ME$$

The materialisation error ME has three components: the error of delimitation DE, of extraction EE, and of preparation PE:

$$ME = DE + EE + PE \text{ whence } TE = IE + DE + EE + PE$$

When all the conditions for correctness are satisfied, in other words when sampling is correct, and only then:

$$DE = EE = PE = 0 \text{ so that } ME = 0 \text{ and } TE = IE$$

It clearly follows that the total sampling error of a correctly taken sample, TE, reduces to the integration error IE, the only error that can be modelled mathematically. It can be concluded that the strategy to be followed regarding materialisation errors ME is:

Do not try to estimate them; eliminate them instead by sampling correctly.

6.3   TAKING A FRACTION OF THE STREAM FOR THE WHOLE OF THE TIME

This method is often used for liquids, sometimes containing suspended solids, in the petrochemical, pharmaceutical and chemical industries. It exists in many forms in which the shape, siting and section of the sampler are variable. There may be times when a simple drawing off from the bottom or side of the delivery pipe is enough. Its principles are shown in Figure 6.10. Although widespread, the method is unfortunately non-probabilistic. It is founded on the hypothesis, rarely realised, of a homogeneous distribution of the constituent elements over the stream's cross-section. Experience shows that it provides only specimens that can be heavily biased, as could be expected from theoretical considerations.

Sampling a fraction of the stream for the whole of the time does not give structurally correct samples and should never be done.

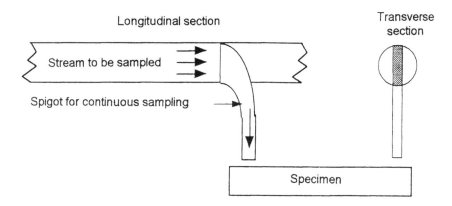

Figure 6.10 Taking a fraction of the stream for the whole of the time

6.4 TAKING A FRACTION OF THE STREAM FOR A FRACTION OF THE TIME

Like the preceding one, this method is also used in the petrochemical, pharmaceutical and chemical industries. The operator may sometimes be satisfied by drawing off a fraction from the bottom of the pipe through a valve actuated by a timer, or again by a simple tap turned on and off at such times as the operator thinks fit, but rarely at regular intervals. Its main points are illustrated in Figure 6.11.

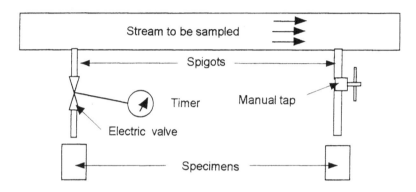

Figure 6.11 Taking a fraction of the stream for a fraction of the time

It is also used for sampling dry solids on a conveyor belt, and is thus represented in Figure 6.1(3). It will be recalled that it was sampling of this nature that proved to be responsible for a relative bias of 9% in the estimated grades of tin ore, as described in Section 1.7.3.

Like the preceding one, this method is non-probabilistic. It, too, is founded implicitly on the unrealistic hypothesis of a homogenous distribution across the stream's cross-section. The method only provides specimens and may be heavily biased, as would be expected from the theory.

Sampling a fraction of the stream for a fraction of the time does not give structurally correct samples and should never be practised.

6.5 THE ONLY CORRECT METHOD FOR SAMPLING FLOWING STREAMS

From the evidence offered in the preceding sections it can be seen that the only probabilistic method for sampling a moving stream is:

Take the whole of the stream for a fraction of the time.

The fraction of the time is shared between a certain number of increments of short duration. The method is easy to apply correctly and the conditions for its fulfilment have been listed above; they are easily put into practice.

The two other methods, taking either a fraction of the stream for the whole of the time or a fraction of the stream for a fraction of the time, are non-probabilistic. They are founded on the hypothesis, usually unrealistic, of a homogeneous distribution of the constituent elements over the stream's cross-section. They always lead to uncontrollable and generally systematic errors and hence must not be used. Their main fault is that they are simple to use and are usually less costly than the correct, probabilistic one. But, only in the short term.

It is, however, necessary to think in the long term, taking into account the financial impact of sampling biases that lead to biased results and to false conclusions, which in their turn lead to non-optimal decision making. Sampling methods that take the whole of the stream for a fraction of the time are the least costly in the long term.

Managers responsible for the control of quality and the general direction of an enterprise must think in the long term and not in the short term. Recalling the words of Etienne Roth [3], given in Chapter 1:

Consequently, at every level concerned with the results of analyses, from the board of directors down to the technician [. . .], there is a need to be aware that the way in which the samples were taken and prepared is just as important as the way in which they were analysed.

Those who establish analytical laboratories must provide both the proper equipment and qualified staff needed to produce representative samples.

Chapter 7

Conclusions on the Qualitative Approach

7.1 SAMPLING IS NOT A SIMPLE TECHNIQUE

Contrary to the belief of the majority of operators, equipment manufacturers, universities and standards committees, sampling is not a simple materials-handling technique. Sampling is a process that generates errors which in the worst cases lead to biases. Sampling is a separate and complete science for which a theory exists. It has been published in stages since 1951 in French (1988) and in English (1992) (see references [6] and [7]), but although it has been received with keen approval by certain English-speaking scientists, it remains neglected, even ignored, on a worldwide scale by universities, by standards committees, and by equipment manufacturers. It is imperative for the analyst and for the person using his or her results that they know how to control (in the strictest sense of the expression) such errors, i.e. how to nullify them where possible, or otherwise how to minimise and then to estimate them.

7.2 HOW CAN THE USER JUDGE A SAMPLING DEVICE?

A user's interest in a sampling device can be expressed in terms of the errors it is likely to make, as follows (and see Section 4.5.2):

- *Accuracy*: defined as the absence of bias or systematic error. It is a property of the mean, which should be zero.
- *Reproducibility*: defined as a low dispersion of the sample values about their mean. It is a property of the variance of the sampling errors, which should be minimal.
- *Representativeness*: defined as a synthesis of accuracy and reproducibility. It

is a property of the mean square, a combination of the mean and the variance of the sampling errors. It is this quality which is sought.

However, the user has no direct method of ensuring that the sample is representative, i.e. that it is accurate and reproducible. It has to be done indirectly. The first guarantee of representativeness is accuracy, which is done by sampling correctly and requires that all the constituent elements of the lot have an equal probability of being selected; correct sampling is always structurally accurate. The second guarantee of representativeness is reproducibility, the quantitative laws for which will be set out in Parts II and III of this book.

7.3   HOW SHOULD A MANUFACTURER DESIGN A SAMPLER?

Designers of a sampler have no direct means of designing an accurate one. They can design a machine or method that may or may not be probabilistic, correct or incorrect. Their first task is to construct a probabilistic sampler. It is a question of design. There is nothing that can be done to improve a non-probabilistic appliance. Its design is structurally wrong. A probabilistic one can be corrected by choosing the correct construction parameters (geometry, cutter speed, etc.). Some incorrect but probabilistic samplers can be corrected, but only at a considerable cost.

7.4   HOW TO USE A CORRECT SAMPLER CORRECTLY

The correctness of a sampler is not an absolute and immutable property. Having carefully chosen a correctly designed and constructed sampler, the user needs to know how to install it, use it and maintain it, all correctly. But the profession of sampling expert is not a self-generating one and whether the personnel involved are in industry, or directly engaged in sample preparation, or are in a laboratory environment, they all need educating.

7.5   EDUCATION, STANDARDISATION AND EQUIPMENT MANUFACTURE

However, no qualified teachers exist, because the two groups able to award or encourage sampling qualifications, i.e. educators and standards committees, remain unaware of the problem. It is just as important to teach sampling theory (hardly ever done in 1998) as it is to teach analysis (very well developed in 1998 and still expanding) because the two sciences/techniques are inseparable and complementary. Sampling and analytical errors are additive and they all contribute to a reduction in the reliability of analyses. But sampling errors are almost invariably far greater than analytical errors.

Thus it is imperative and urgent to organise the teaching of sampling at two levels: one for the engineer and one for the higher-grade technician. It is equally imperative and urgent to reform sampling standards. There are national and international standards committees dealing with this subject that are, at best, simplistic or silent. At worst, they continue to make wrong and dangerous recommendations that are written in full ignorance of the theory. Analytical standards remain content to state that 'analyses must be done on representative samples' without defining this concept, any more than they define the equally important concept of correctness.

In the absence of any serious education or standardisation, no makers of reliable sampling equipment are to be found. Why should they pay more attention to the theory than do the educators and standard setters? The author has himself at times had to design and supervise the construction of the samplers needed by his clients.

7.6 PRACTICAL CONCLUSIONS

Only a correctly designed sampler can produce samples that are representative and hence reliable.

This recommendation will continue to remain a dead letter as long as the two principal branches of training and regulation, i.e. of education and standardisation, continue to ignore sampling theory and, above all, the concepts of correctness and representativeness. Whence follows the second recommendation, a corollary of the first:

> *It is imperative and urgent that the theory of sampling is brought into education and standardisation. And, before that can be done, how are the teachers and regulators to be trained?*

Part II

THE QUANTITATIVE APPROACH

TAKING REPRODUCIBLE SAMPLES FROM A POPULATION OF NON-ORDERED OBJECTS

Preface to Parts II and III

The lot to be sampled can be treated as a zero- or one-dimensional object, and the aim is to take a 'representative' sample from it. It has been shown that ensuring representativeness requires the attainment of two objectives that are independent of each other: 'accuracy' and 'reproducibility'. In other words, we have to answer the two questions 'how' and 'how much'?

Part I answered the question 'how?'. We now know what must either be done or avoided to ensure that the sample is structurally accurate. We know that sampling is accurate if and only if it is done correctly. Correctness is thus a necessary condition. The conditions for correct sampling have been defined above. They are simple and easy to implement.

There is not, and there cannot be, a non-probabilistic sampling theory, and even if a probabilistic sampling theory of unequal probability has been developed (Chapter 17 of reference [7]), such a theory offers no practical solution if sampling is incorrect because it assumes that the probabilities P_i of taking the constituent elements F_i are known in advance, and they never are.

There is thus no point in trying to answer the question 'how much?' when sampling is non-correct, i.e. non-probabilistic or probabilistic but incorrect, because we know that whatever its mass a sample taken in these conditions will be structurally biased to an extent impossible to foresee. Inevitably, it will be non-representative.

That is why, when embarking on Parts II and III and preparing to answer the question 'how much?' we assume that everything has been done to ensure that sampling will be correct and therefore accurate: it will be assumed explicitly that all the constituent elements of the lot L have an equal probability P of being selected. We can thus redefine the title of these last two parts by summing them as: *ensuring the reproducibility of a correct sampling process.*

- Classical or non-weighted statistics are concerned with populations of things, material or non-material, that have equal statistical weight. They are not assumed to be ordered—even if some order existed it would not be taken into account. The results of repeated chemical analyses made under the same conditions are a good example. Opinion polls are another. *A priori*, each result, each opinion, is of equal value and hence has the same statistical weight. However, non-weighted statistics are not applicable to the problems needing to be solved here. In the quantitative development of an appropriate sampling theory one needs to distinguish between two very different cases that have not been studied by classical, non-weighted statistics.
- 'Weighted' statistics are concerned with populations of objects with different masses (for example, solid mineral grains with masses in ratios up to $1:10^{20}$) to which it is necessary to assign proportional statistical weights. The objects are not assumed to be ordered—even if some order were present it would be disregarded.
- 'Chronostatistics' are the statistics of series of time-based variables. They deal with variables, material or non-material, of equal or different masses, that are ordered and whose properties are potentially autocorrelated.

Taking correct and reproducible samples from populations of non-ordered objects of equal or different masses is dealt with in Part II (Chapters 8 to 10). Taking correct and reproducible samples from time-based, ordered variables of equal or different masses is dealt with in Part III (Chapters 11 and 12).

Chapter 8

The Heterogeneity of a Population

To understand and know, it is first necessary to distinguish [...] To distinguish is to designate. (Jean d'Ormesson, *Almost nothing about almost everything.*

8.1 A REMINDER ON HETEROGENEITY AND SAMPLING

It was seen in Chapter 3 that without exception all sampling errors are the direct consequence of one form or another of heterogeneity. It is therefore natural, when making a quantitative study of sampling errors, to first define, distinguish and designate as suggested by d'Ormesson, and to quantify each form of heterogeneity (Chapter 8); and then in the second place (Chapters 9 and 10) to deduce the mean and variance of each of the two components of the sampling error in zero dimension by the simple application of a probability function.

8.2 QUANTIFYING POPULATION HETEROGENEITY

Readers have been promised two things that appear to partly contradict each other: first, that this book will not be encumbered by heavy mathematical proofs, but second, that the processes generating sampling errors will be made clear to them.

As always in such cases a compromise has to be made which, it is hoped, will prove to be acceptable. Indeed, a mathematical proof is a powerful tool for the analysis and understanding of any physical phenomenon. To neglect it would not seem reasonable.

8.2.1 The Contribution Made to the Heterogeneity of Lot L by Some Unspecified Unit U_m

Lot L, the subject of the study, can be regarded in general terms as a population of unspecified units U_m that can be defined later as the case requires. A is the 'critical component' or analyte, the component of interest. References to the grade or concentration of component A are to its actual but unknown grade, not to some estimate of it. The following definitions are made:

N_U number of units U_m in lot L

M_m mass of the unit U_m

M_L mass of lot L. It is the sum $\sum_m$ of the masses M_m of the N_U units U_m

M_{m*} mass of the mean unit U_{m*} of lot L, defined as the mean of the masses M_m

a_m the grade of the analyte A in unit U_m (i.e. the proportion by mass of A in U_m)

a_L the grade of analyte A in lot L (i.e. the proportion by mass of A in L)

h_m the contribution to the heterogeneity of lot L made by U_m, defined as follows:

$$h_m = \frac{(a_m - a_L)}{a_L} \cdot \frac{M_m}{M_{m*}} \quad \text{with } a_L = \frac{\sum_m a_m M_m}{\sum_m M_m} \quad \text{and } M_{m*} = \frac{M_L}{N_U}$$

Note 1: The grade a_L is a weighted mean. It follows that any contribution to heterogeneity must be weighted accordingly. Attention had previously been drawn to the function h_m when formulating the equiprobable model in 1951. The variance of equi-probable sampling was in fact proportional to the variance of h_m within the population of the N_U units (the fragments) U_m. h_m is the starting point from which 'constitutional and distributional heterogeneities', which play a major role in sampling theory, will be defined and quantified.

Note 2: In classical statistics the masses of the units are not usually taken into account and they are regarded as being uniform. The non-weighted heterogeneity h'_m can thus be written: $h'_m = (a_m - a_L)/a_L$. It is the relative deviation of a_m about its mean a_L. The contribution h''_m to the heterogeneity of L made by the masses of the units can be defined similarly:

$$h''_m = (M_m - M_{m*})/M_{m*}$$

An instructive example is given in Section 11.7.3 (Figure 11.6).

8.2.2 The Constitutional Heterogeneity CH_L of Lot L: $U_m = F_i$

Initially, the scale of observation will be such that unit U_m is a fragment, a molecule or an ion F_i. The following are thus defined:

h_i the contribution of the fragment F_i to the heterogeneity of L

$$h_i = \frac{(a_i - a_L)}{a_L} \cdot \frac{M_i}{M_{i*}} \text{ with } i = 1, 2, \ldots N_F \text{ and } M_{i*} = M_L/N_F$$

CH_L the constitutional heterogeneity of lot L defined as the variance of h_i:
$CH_L = s^2(h_i)$, noting that $m(h_i) = 0$

As long as the fragments, molecules, ions F_i remain unaltered the constitutional heterogeneity CH_L is an intrinsic property of lot L.

8.2.3 Distributional Heterogeneity DH_L of Lot L: $U_m = I_n$

The scale of observation is now that at which the unit U_m is a group I_n of neighbouring fragments, molecules or ions. The following can be defined:

h_n the contribution of the group I_n to the heterogeneity of L, defined as:

$$h_n = \frac{(a_n - a_L)}{a_L} \cdot \frac{M_n}{M_{n*}} \text{ with } n = 1, 2, \ldots N_I \text{ and } M_{n*} = M_L/N_I$$

DH_L the distributional heterogeneity of the fragments F_i within L, defined as:
$DH_L = s^2(h_n)$, noting that $m(h_n) = 0$

Unlike the fragments F_i, the groups I_n of fragments can alter. Any mixing, homogenising or segregation alters the distribution of the fragments between the groups and thus the value of DH_L. *Mixing and homogenising reduce DH_L; segregation increases it.*

8.2.4 The Properties of CH_L and DH_L: Relationships with Sampling

As a sample E is made up of the union of the Q increments I_q, which are the I_n groups selected, the sampling variance is proportional to DH_L, which is itself proportional to CH_L. It has been shown (Section 4.4.15 of reference [7]) that:

$$DH_L = CH_L \cdot \frac{1 + YZ}{1 + Y} \text{ (dimensionless)}$$

in which:

Y is a dimensionless grouping parameter characterising the size of the increments I_q. Y is positive or zero: $Y >= 0$, with:

  $Y = 0$: when the increments consist of a single constituent element, and only then.

Z is a dimensionless segregation parameter characterising the type of distribution of the constituent elements within lot L, and $1 >= Z >= 0$ with:

  $Z = 1$: when their distribution is completely segregated

  $Z = 0$: when the material is completely homogenised (which does not mean that it is 'homogeneous'). The distribution is then characterised by there being no correlation between the positions of the elements and their physical or chemical nature.

8.2.5   The Heterogeneity Invariant HI_L

The parameter CH_L can very rarely be estimated. It is used only in the development of the theory. In practical applications it is replaced by:

HI_L The heterogeneity invariant defined below, which can always be calculated, at least to its order of magnitude:

$$HI_L = CH_L \cdot \frac{M_L}{N_F} \quad \text{(dimensions of mass in grams)}$$

8.3   THE CHARACTERISTIC PARAMETERS OF A SUBSTANCE

8.3.1   Analysis of HI_L for Solid Fragments

In the case of solid fragments the heterogeneity invariant HI_L can be expressed as the product of five factors, called the descriptive parameters (Section 4.9 of reference [7]):

$$HI_L = c\beta f g d^3 \text{ (dimensions of mass in grams)}$$

which are defined as follows:

c the constitution parameter, having the dimensions of density expressed in g.cm^{-3}. It accounts for the densities and the proportions of the constituents of the lot. It is defined in Section 8.3.2 below.

β the liberation parameter (dimensionless), accounting for the degree to which the critical substance is embedded in the other fragments (Section 8.3.3)

f the shape factor (dimensionless) (Section 8.3.4)

g the size range factor (dimensionless) (Section 8.3.5)

d parameter describing the size of the coarsest fragments (defined as the size of the sieve aperture that retains 5% of the mass of the lot. It must be expressed in cm so as to be consistent with the dimensions of the other parameters) (Section 8.3.6).

These parameters can always be calculated, at least to their order of magnitude.

8.3.2   The Constitution Parameter c

This is defined mathematically, for a material consisting of two components, as:

$$c = \frac{(1 - a_L)}{a_L} \cdot [(1 - a_L)\delta_A + a_L \delta_G](\text{g.cm}^{-3})$$

in which:

a_L is the decimal proportion of component A in lot L (e.g. 10% = 0.10)

δ_A is the density of the critical component A (g.cm^{-3})

δ_G is the mean density of the remaining, non-critical, components. The subscript G, the initial letter of the word ' gangue', follows from the mineral origins of the theory. In practice the bracket:

$$[(1 - a_L)\delta_A + a_L\delta_G]$$

has a value close to the mean density δ_M, and an allowable approximation is:

$$c \sim \delta_M \frac{(1 - a_L)}{a_L} \quad \text{or if } a_L << 1 \text{ then } c = \frac{\delta_M}{a_L}$$

The quotient $(1 - a_L)/a_L$ varies from unity if $a_L = 0.5$, to 10^9 if $a_L = 1$ppb, for example.

It can be seen that as the maximum value of density is only of the order of a few units at the most, the major factor determining the value of c is the grade a_L, and d^3 will act similarly on the value of HI_L.

8.3.3 The Dimensionless Liberation Parameter β (Solids)

This parameter takes its name from mineralogical materials in which several minerals occur in the same 'constituent element' (the fragment). However, the same conditions can be seen in the recovery of scrap metal from old cars or electrical installations; indeed, from waste of all kinds. Examples are copper wires sheathed in insulation; scrap from the electronics industry; bottle banks in which the glass is contaminated with metal caps and labels.

The value of β varies from unity to zero. The value $\beta = 1$ is reached when the components are completely separated from each other, or 'liberated' in the terminology of minerals engineering. The value $\beta = 0$ occurs when the components are completely homogenised, an unattainable situation in practice. There is no simple mathematical formula giving the value of β. When unsure, it is best to set $\beta = 1$.

In previous work by the author he proposed a rule of thumb whereby, in the absence of any other information, $\beta = (d_{\text{lib}}/d)^{0.5}$ in which d_{lib} is the liberation diameter. However, François-Bongarçon proposes (in a paper to be published shortly) that, based on the results of his experience, in the case of gold ores:

$$\beta = (d_{\text{lib}}/d)^{1.5}$$

8.3.4 The Dimensionless Shape Parameter f (Solids)

This is the fraction of the actual volume of the fragment to that of a cube whose side has the same length as the sieve aperture that allows the fragment to pass through it. It can thus be defined as a coefficient of 'cubicity'. For a sphere f is

0.52. Most compact fragments have an f-value near to 0.5, and suitable f-values for other shapes are easy enough to calculate.

8.3.5   The Dimensionless Size Range Parameter g

The value of g depends on the definition of d (Section 8.3.6 below or Chapter 5 of reference [16]). The following values are used in practice:

Undifferentiated, unsized materials. Mean value	$g = 0.25$
Undersize material passing through a screen	$g = 0.40$
Oversize material retained by a screen	$g = 0.50$
Material sized between the two screens	$g = 0.6/0.75$
Naturally sized materials, e.g. cereal grains	$g = 0.75$
Uniformly sized objects, e.g. bearing balls	$g = 1.0$

8.3.6   The 'Size' or 'Diameter' of the Largest Fragments d (cm)

The 'size' or 'diameter' of the largest fragments of a lot is the size of the aperture of a square-mesh screen that retains 5% of the material and passes 95% of it; this size is denoted by d_{95} or simply by d. The definition has a physical meaning only in the case of solid fragments. For molecules and ions one has to proceed by analogy and this is not always easy when dealing with organic molecules. Because the diameter d is raised to the third power in the expression for HI_L it needs to be evaluated as precisely as possible.

The same remark applies to the critical content a_L which determines the order of magnitude of the heterogeneity invariant HI_L, and hence that of the sampling variance too.

Chapter 9

Sampling: The Zero-dimensional Model: The Fundamental Error FE

9.1 DEFINITION OF THE FUNDAMENTAL ERROR FE

This is the total sampling error made when the N_F elements of lot L are submitted to a selection process that is:

- Correct, i.e. probabilistic and with a uniform probability P of being selected, but, in addition,
- The elements are selected independently of each other, that is, one by one, sequentially.

This model is equivalent (although described differently) to the 'equi-probable model' of 1951–3 ([1, 2]) when it was shown that, among all the components making up the total sampling error TE, FE is the only error that cannot be reduced to zero, even if the most favourable hypotheses are accepted. FE is therefore the minimum, irreducible sampling error, and so justifies the title of 'fundamental error' given to it.

One neither knows how, nor is able, to do better.

9.2 THE DISTRIBUTION LAW OF FE

It is impossible to understand certain properties of the fundamental error without making at least a modest mathematical effort. A reader not interested in doing so may go straight to the next section.

The grade a_E of a sample is defined as the quotient of two random variables: the numerator is the mass A_E of the critical component A in the sample E: the denominator is the mass M_E of the relevant components (i.e. those involved in the definition of the grade, for example dry solids) in the same sample. In general, this quotient follows no simple distributional law.

Nevertheless, certain writers have shown (see Chapter 17 of reference [7]) that the distribution law of a_E is approximately normal when the two following conditions are satisfied simultaneously:

1. Both numerator and denominator are at least approximately normal
2. Both numerator and denominator have low coefficients of variation (= the quotient of the standard deviation over the mean).

In general, both the above conditions are satisfied if two others hold good:

3. The mean grade a_L of lot L is not too small (which eliminates trace elements, e.g. in which $a_L < 1$ ppm).
4. The uniform probability of selection P is not too low.

In fact conditions 3 and 4 reduce to a single one: the number of critical elements in the sample should not be 'too small'. Whence follows the eternal question put to statisticians and other specialists in probability theory: what is a 'large number'? For practical purposes it will be assumed here that a 'too small number' is less than 30, and a 'large number' is at least 50.

If $a_L = 1$ ppm, the mathematical expectation of a_E is also about 1 ppm. Assuming that the components, critical or otherwise, have similar masses and are completely liberated, the sample will contain, on average, one critical element in one million. If the sample has to contain, on average, at least 30, or better 50, critical elements, the average mass of sample $M_E = PM_L$ must be at least equal to 50 million times that of the average fragment, from which M_E can be found and knowing the mass M_L, the probability P below which it would be dangerous to proceed can be determined. It is a simple matter to extend the same calculations to ultra-trace elements in concentrations of 1 ppb, 1 ppt, etc.

Practical example

An alluvial deposit comprises liberated gold at 1 g/t (i.e. 1 ppm) in 0.2 mm quartz grains (the gold grains have the same mean diameter as the quartz grains). The specific gravity of gold is 19.3. A single spherical gold grain 0.2 mm diameter has a volume of 0.004 mm^3 and weighs about 0.076 mg. A 1 kg sample contains, on average, 1 mg or 13 grains of gold. Obviously, this number is too small. If one adheres to the rule of 50 fragments, a sample will contain, on average, 3.8 mg gold and weigh about 3.8 kg. It follows, for example, that if M_L is 100 kg, the uniform probability $P = 0.04$ (4%).

As soon as fewer than 30 to 50 critical particles are gathered into the sample the distribution of their grades becomes increasingly less normal, showing increasing skewness and tending towards a 'log-normal' distribution. One of the sampling characteristics of the lognormal distribution is the existence of a structural bias inherent in the law of 'log-normality'. It is too large to be neglected and it is an inevitable consequence.

9.3 THE MEAN *m*(FE) OF THE FUNDAMENTAL ERROR FE

Surprising as it may seem since one can do no better, the mean m(FE) of the distribution of the fundamental error FE is never strictly structurally zero even if it is completely negligible (Table 17.1 in reference [7]). This property is neither the result of an error in the development of the model, which has been checked by various mathematicians, nor of some maladroitness by its author. It is a result of the fact that the grade of a sample is the quotient of two random variables that can be expressed as a power series, of which only the first term is equal to a_L (Section 17.5 of reference [7]). The mean is structurally reduced to zero only if one of the two following conditions is satisfied (a fact serving to test the consistency of the model):

- $P = 1$: the sample is the lot itself
- $a_i = a_L$ for all i: the lot is completely homogeneous.

Both conditions are unrealistic. The mathematical expression for the first term of the bias is given in Chapter 19 of reference [7], and the reader is referred to this.

9.4 THE VARIANCE s^2(FE) OF THE FUNDAMENTAL ERROR FE

It has been shown (Chapter 19 of reference [7]) that the variance s^2(FE), referred to for convenience as the 'fundamental variance', is proportional to the two parameters CH_L and HI_L that characterise the constitutional heterogeneity:

$$s^2(\text{FE}) = \frac{1-P}{P} \cdot \frac{CH_L}{N_F} = \left[\frac{1}{M_E} - \frac{1}{M_L} \right] HI_L \sim \frac{c\beta fgd^3}{M_E} = \frac{Cd^3}{M_E}$$

The first equality is the result of a theoretical study; the second of replacing CH_L by its expression in terms of HI_L. The penultimate equality is only approximate (see section 4.7 of reference [7]). Moreover, it assumes that the mass of the sample is negligible compared with that of the lot, which is often the case.

The final expression underlines the importance of the two main factors (d^3 and M_E), and that the product of the four factors $c\beta fg$ can for all practical purposes be expressed as a single constant C. In this form it allows the fundamental variance to be calculated, at least to its order of magnitude, bearing in mind that it is the ultimate minimum value of the total sampling variance.

9.5   CONDITIONS FOR NULLIFYING THE VARIANCE $s^2(\text{EF})$

From the two previous expressions for the variance it is evident that it reduces to zero if $P = 1$ or if $M_E = M_L$, which gives some assurance about the formula's validity. HI_L, the product of five factors, becomes zero if any one of them is zero. From its definition the constitution parameter c is itself the product of a fraction and a bracket. The bracket is the sum of two positive or zero terms that cannot both be zero at the same time. It is therefore never zero. The other factors are either strictly positive or only reduce to zero at the limit if unrealistic hypotheses are set up.

   The fraction $(1 - a_L)/a_L$ only becomes zero if $a_L = 1$, that is, if the material is composed of one completely pure constituent, a totally unrealistic condition. It therefore follows that:

   the fundamental variance $s^2(FE)$ is never zero.

9.6   CONDITIONS FOR MINIMISING THE VARIANCE $s^2(\text{FE})$

From the last two expressions for the variance it follows that the parameters c, β, f and g are factors fixed by the nature of the material in the lot. On the other hand, at least to a certain extent, d and M_E can be changed. In the case of solid fragments d can be reduced, so that sample-preparation procedures can be planned as a series of alternating size and mass reductions.

   The second solution, obviously, is to increase the mass M_E of the sample. The disadvantage of these two solutions is their increased cost. This is why a sample preparation procedure is often, if not always, a compromise between a higher cost and a gain in reproducibility. It is important to accept from now on that sampling reproducibility involves a cost. In fact:

   There is no such thing as an unsolvable sampling problem. There are only solutions that are judged, sometimes wrongly, to be too costly.

9.7   SOLUTION OF PROBLEMS INVOLVING THE VARIANCE $s^2(\text{FE})$

Having characterised a particular material by its sampling constant C, which can usually be estimated, the sampling formula can be used to solve three kinds of problem:

1. To calculate the fundamental variance $s^2(\text{FE})$ given d and M_E

2. To calculate the minimum mass M_{L0} of sample to take, given d and a maximum allowable variance s_0^2 that must not be exceeded
3. To calculate d given fixed values of M_E and s_0^2. That is, to what size the lot must be crushed or pulverised before sampling if the sample mass is M_E and the variance has to be less than s_0^2:

$$s^2(\text{FE}) = Cd^3/M_E \qquad M_{E0} = Cd^3/s_0^2(\text{FE}) \qquad d^3 = Cs_0^2(\text{FE})/M_E$$

9.8 CALCULATING THE CONFIDENCE LIMITS $\pm 2s(a_E)$

In this expression $s(a_E)$ is the standard deviation of the distribution of the grade a_E of the sample. Calculating $s(\text{FE})$ and $\pm 2s(\text{FE})$ are easy enough once $s^2(\text{FE})$ is known. In calculating $s(a_E)$ it must be remembered that FE is a relative error and that:

$$s^2(\text{FE}) = s^2(a_E)/a_E^2 \qquad \text{whence } s(a_E) = a_E s(\text{FE})$$

Example

If $s^2(\text{FE}) = 4.10^{-4}$ then $s(\text{FE}) = 2.10^{-2}$.

If the grade a_E is, for example, 10% A (or 0.10), it is easy enough to show that:

$$s(a_E) = a_E s(\text{FE}) = 0.1 \times 2.10^{-2} = 2.10^{-3} = 0.2\% A$$

Assuming a normal distribution, the 95% confidence limits for the grade a_L are:

$$a_L = a_E \pm 2s(a_E) = (10\% \pm 0.4\%)A$$

9.9 RESUMÉ OF THE PROPERTIES OF THE FUNDAMENTAL MODEL

This model is based on the assumption that all the constituent elements of the lot to be sampled have an equal probability of being drawn in a selection process that takes the elements one by one and independently of each other. It is known that, in these conditions:

1. On the one hand, sampling is (in practical terms) unbiased. The mean error is negligible, except in the limiting case of trace elements (see Section 9.3 above).
2. On the other hand, it assures maximum reproducibility, i.e. minimum variance.

3. As a corollary of 1 and 2 the mean squared error is minimum, thus ensuring maximum representativeness.

It is therefore concluded that:

It is not possible to sample more effectively.

Chapter 10

Sampling: The Zero-dimensional Model: The Segregation and Grouping Error SGE

10.1 DEFINITION OF THE MINIMUM PRACTICAL ERROR MPE

Following the presentation of the equi-probable model in 1951 [1] an experiment was made, as described in 1953 [3], which had as its object a comparison of the variance actually observed with that forecast by the model. It confirmed that the actual variance is greater than the predicted variance. From this it was concluded that real sampling conditions depart from the ideal ones of the model and they introduce a supplementry error that has to be explained. It was not until 1975 that this problem was solved [17], after a detailed study of the concept of heterogeneity.

Under the ideal conditions used in the model, i.e. when the 'units' of the increments are 'individual fragments', it is the constitutional heterogeneity CH_L which is involved and the total error variance so generated is proportional to it. This error, which is known to be the irreducible minimum, has been called the 'fundamental error FE'. To summarise:

Increments	Type of heterogeneity involved	Error
Individual fragments F_i	Constitutional heterogeneity CH_L	Fundamental error FE

Under practical selection conditions, i.e. when the 'units' acting as 'increments' are 'groups of neighbouring fragments I_n' and are no longer 'individual fragments F_i', then, by definition, the distributional heterogeneity DH_L is involved and the total sampling error variance so generated is proportional to it. This error is called the 'minimum practical error MPE'. To summarise in this case:

Increments	Type of heterogeneity involved	Error
Groups of fragments I_n	Distributional heterogeneity DH_L	Minimum practical error MPE

Because of the possible existence of some correlation between neighbouring fragments—most often because of gravitational effects—the fragments in the increments are no longer independent of each other. It is this loss of independence which generates the supplementary error.

10.2   THE VARIANCE s^2 (MPE) OF THE MINIMUM PRACTICAL ERROR MPE

By comparing s^2(MPE) with the variance s^2(FE) defined in Section 9.4, and observing the symmetry of the two variances:

$$s^2(\text{MPE}) = \frac{1-P}{P} \cdot \frac{DH_L}{N_I} \quad \text{and} \quad s^2(\text{FE}) = \frac{1-P}{P} \cdot \frac{CH_L}{N_F}$$

with m(MPE) ~ 0 and m(FE) ~ 0.
From Section 8.2.4:

$$DH_L = CH_L \cdot \frac{1+YZ}{1+Y} \quad \text{(dimensionless)} \quad \text{in which}$$

Z is a dimensionless segregation parameter characterising the distribution of the fragments within lot L. Z lies between 0 and +1: $0 <= Z <= 1$.
Y is a dimensionless grouping parameter characterising the size of the increments. Y is positive or zero: $0 <= Y$.

Furthermore, it has been shown in Section 4.4.11 of references [6] and [7] that $(1 + Y)$ was, from the definition of Y, equal to $(N_F - 1)/(N_L - 1)$, from which it can thus be deduced that:

$$\frac{s^2(\text{MPE})}{s^2(\text{FE})} = \frac{DH_L}{CH_L} \cdot \frac{N_F}{N_I} = (1+YZ)\frac{N_F(N_I - 1)}{N_I(N_F - 1)}$$

As N_F and $N_I \gg 1$, the last quotient is virtually equal to 1. Also, it will be remembered, the product YZ is always positive or zero: $YZ >= 0$, and hence:

$$s^2(\text{MPE}) = (1 + YZ)s^2(\text{FE}) >= s^2(\text{FE})$$

10.3 DEFINITION OF THE SEGREGATION AND GROUPING ERROR SGE

The 'segregation and grouping error SGE' is defined by the following equalities:

$$MPE = FE + SGE \quad \text{and} \quad s^2(MPE) = s^2(FE) + s^2(SGE)$$

As its name indicates, and as will be demonstrated shortly, the segregation and grouping error SGE arises from the conjuction of two factors neither of which, in the absence of the other, generates an error.

10.4 THE MEAN OF THE SEGREGATION AND GROUPING ERROR SGE

The mean, as with those of MPE and FE, is virtually zero.

10.5 THE VARIANCE OF THE SEGREGATION AND GROUPING ERROR SGE

The variances $s^2(MPE)$ and $s^2(SGE)$ can be written:

$$s^2(MPE) = (1 + YZ)s^2(FE) = s^2(FE) + YZs^2(FE)$$

from which it follows that:

$$s^2(SGE) = YZs^2(FE)$$

10.6 NULLIFYING THE SEGREGATION AND GROUPING ERROR SGE

This requires the cancellation of the variance, which is the product of three factors.

10.6.1 Nullifying the Fundamental Variance $s^2(FE)$

It was shown in Section 9.5 that:

The variance s² (FE) never cancels and is always positive.

10.6.2 Nullifying the Grouping Parameter Y

Section 8.2.4 demonstrated that the grouping parameter Y cancels out in the singular case that the increments are 'mono-fragmentary', i.e. each consists of one and only one fragment. Such a solution is economically impracticable except in experimental studies.

The grouping parameter Y never cancels and is always positive

10.6.3   Nullifying the Segregation Parameter *Z*

It was seen in Section 8.2.4 that the segregation parameter *Z* cancels out in the limiting case when the fragments of the lot have been 'perfectly homogenised', and only then.

The segregation parameter Z cancels out only at such a limit, mainly in the case of liquids or very fine powders. Otherwise, it is always positive or zero.

10.6.4   Conclusions Concerning the Nullification of s^2(SGE)

Three main cases must be considered:

1. *Solids*: There is no practical way of cancelling the segregation and grouping error SGE, except in the case where the lot is of low weight, consists of dry, fine powder, and the density contrast between its constituent elements is low. In every other case homogenising is incomplete, random and costly.
2. *Liquids*: When the lot can be handled with relative ease, agitation and mixing, whether by hand, mechanically or magnetically, is enough to ensure sufficient homogenisation, except with liquids of high viscosity. In every other case, as with solid fragments, homogenising is incomplete, random and costly.
3. *Multi-phase mixtures*: With mixtures of liquids and fine solid particles, e.g. those usually found in mineral processing, there is always a high contrast between the densities of the liquids and the solids. The liquid is usually water. (The less dense minerals are coal products with densities of the order of 1.4; schists, carbonates, silica and silicates have densities in the range of 2.2 to 2.7.) There is always the risk of spontaneous segregation, slightly less so if the liquid is viscous. Violent agitation brings about a certain degree of homogenisation, but it is usually unstable; the solids begin to segregate as soon as agitation ceases. Some degree of stable homogenisation is obtained only in the case of suspensions that are thick and concentrated, and in which the solids are very fine and well dispersed.

With mixtures of gases and fine solid particles, such as the raw materials and crushed clinker found in the cement industry, the contrast between the densities of about 2.5 for the solids, on the one hand, and air, on the other, makes homogenisation practically impossible. This is one of the reasons why sampling pneumatically transported solids is so difficult to do correctly.

10.7 MINIMISING THE VARIANCE OF THE SEGREGATION AND GROUPING ERROR SGE

Although it is difficult to nullify it completely, a re-examination of the preceding paragraphs will show how the variance of the error SGE can be minimised.

- The fundamental variance $s^2(FE)$. Minimising this should be the prime objective of any sampling process.
- The grouping parameter Y. To minimise Y it is best to reduce the size of the increments as far as this is possible. If the mass of the sample is fixed, for example with the intention of minimising the fundamental variance at acceptable cost, it is better to take a large number of small increments than a small number of large ones. This has the further advantage of reducing the integration error IE that was described in Section 5.2.6 and which will be studied in Part III.
- The segregation parameter Z. To minimise Z it is best to reduce the distributional heterogeneity of the lot as far as possible by 'homogenising' it before sampling, by agitation, mixing, etc., manually or mechanically. This can only be done easily with manageable lots. Laboratory samples fall into this category, a fact that analysts and makers of laboratory equipment ought not to lose sight of, as related in Section 4.10.2. The makers should emphasise that their equipment gives reliable and unbiased, i.e. significant, results only with materials that have been thoroughly homogenised.

10.8 HOMOGENISATION AND BED-BLENDING

The technique known as 'bed-blending' is used only on an industrial scale, particularly when dealing with the raw materials of cement works. Chapters 35 of references [6] and [7] describe the theory and the practice of the technique. In practice, the distributional heterogeneity is reduced only along a horizontal axis. The distribution remains perfectly heterogeneous, or even segregated, in the other two dimensions of the heap. Bed-blending achieves only a 'unidimensional homogenisation', although that by itself may be considerable. 'Bed-blending' usually precedes grinding. The latter alone achieves three-dimensional blending; an example of this is the ground mixture fed to a cement kiln.

The author has always believed that industry, which tends to regard bed-blending as a very onerous and therefore very costly technique, has not taken advantage of the process to the extent claimed by its supporters. However, there is nothing to prevent its use at the scale of a pilot plant or even in the laboratory, where it has given rise to the inexpensive technique of the 'sampling rail' invented by J.-M. Pujade-Renaud. Bed-blending allows, providing certain precautions are taken (see Chapter 35 of references [6] and [7]), a processing plant to be fed with material of almost uniform composition, and this always gives the maximum yield.

10.9 AN INSTRUCTIVE EXAMPLE DRAWN FROM CARD GAMES

Consider the deal from a hand of bridge in the light of the above definitions. The dealer carefully shuffles the pack, the 'lot' of 52 cards, and cuts it before dealing them one by one into four hands in the manner of true fractional shovelling described in Section 2.2.2. Each one of the four hands is a 'sample' with a sampling rate of $1/N = 1/4$ of the lot of 52 cards. How can this procedure be analysed logically?

When shuffling the pack the dealer is trying (without being conscious of the fact) to destroy any correlation—which must have arisen during the playing of the preceding hand—between the identity of each of the 52 cards and its position within the pack. The destruction is progressive, asymptotic and random. If it is assumed that the cards have been shuffled perfectly, the pack is then completely homogenised and the segregation parameter Z is cancelled. But shuffling is never perfect. The parameter Z characterising the distribution only reduces to zero asymptotically and randomly. Furthermore, the way in which the cards are dealt adds a second precaution: the cards must be dealt one by one. This ensures that the grouping parameter Y is cancelled, and with it the product YZ and the variance s^2 (SGE). The object of cutting the cards is to ensure that the hands are dealt randomly to the four players, without favouring anyone. This guarantees that the game proceeds equitably.

First, the segregation parameter Z is reduced by shuffling the pack as well as possible, and then the grouping parameter Y is cancelled by dealing the cards one by one. Note that the decisive factor in cancelling the product YZ is that the cards are dealt one by one. The preliminary shuffle has the effect of partially suppressing the cyclic grouping of period four which is related to the fact that, in the 13 tricks taken in the previous hand, there are always several containing four cards of the same suit. The four hands of 13 cards dealt to the four players are samples with maximum representativeness as the segregation and grouping error SGE has been completely cancelled, and the residual fluctuations between the four hands (which provide the game's principal interest) are reduced to the ultimate minimum represented by the fundamental error FE. One may pause and reflect that shuffling tends to reduce the contrast between the four hands and brings about 'average' hands at the expense of more extreme ones.

Unlike bridge the game of belote, very popular in France, is played with a pack of 32 cards (if the rules of the game have not changed in the past 60 years) and hands of eight cards are dealt, after shuffling, in two groups of three cards each and one of two cards. While shuffling has the same effect as in bridge in reducing the parameter Z, dealing in groups of several cards each introduces a segregation and grouping error SGE because the grouping parameter Y no longer cancels.

If a numerical value is given to each card, and if the value of a hand is equal to the sum of the card values in it, a simple experiment would show (given an equal number of cards in each hand, 13 for bridge, 8 for belot) that hands dealt in groups of two or three cards show a greater variance in their values than hands in which the cards have been dealt one by one.

10.10   HYPERHOMOGENEITY

The reader may be suprised to see that homogenising lots of material and shuffling packs of cards does not cancel the distributional heterogeneity, but can only reduce the mean of its distribution, which remains random. A zero value for the distributional heterogeneity cannot be attained by means of a homogenising action that is always random. It is only possible by means of an intelligent and therefore deterministic distribution of fragments, molecules and ions (which is unrealistic), or in the case of cards by them being dealt in the way the author's grandchildren did when young, i.e. laying the cards out face upwards and giving each player one and only one ace, one and only one king, and so on.

This is the domain of an 'artificial and intelligent' hyperhomogenisation, as distinct from a 'natural and random' one, in which the segregation parameter Z takes a negative value, a fact underlining its artificial nature.

Nevertheless there does exist in nature a kind of spontaneous hyperhomogeneity, and it is the modular homogeneity found in perfect crystals that follow highly deterministic forms.

10.11   STRATEGY REGARDING THE SEGREGATION AND GROUPING ERROR SGE

The segregation and grouping error is rarely actually determined. If it were needed, the variance of the minimum practical error MPE would be found by experiment and then the variance of the fundamental error FE by calculation. The variance of SGE is their difference.

More often it is sufficient, having regard to the nature of the problem, to minimise the SGE as demonstrated above.

The segregation and grouping error SGE is one of the errors that one seeks to minimise, for want of being able to cancel it.

PART III

THE QUANTITATIVE APPROACH

ENSURING THE REPRODUCIBILITY OF SAMPLES TAKEN FROM A SERIES OF CHRONOLOGICALLY ORDERED OBJECTS

Chapter 11

Heterogeneity: The One-dimensional Model

11.1 REVIEW OF THE CONDITIONS FOR THE MODEL TO BE VALID

It is useful to recall that in the preface to Parts II and III it was stated explicitly that the models developed and presented here presume that sampling is correct and therefore accurate.

11.2 LINKING THE ZERO- AND ONE-DIMENSIONAL MODELS

These two models, which represent a single phenomenon, are complementary and non-contradictory. In fact they occur together, one in the other, at the small scale of the increments. The one-dimensional model has been adapted to moving streams of material in which the dimension corresponding to the length is very much greater than the two dimensions of the transverse section. For example, a moving lot L can be considered to comprise the 32 potential increments shown in the upper part of Figure 11.1.

A lot that has travelled for one hour on a belt moving at 2 m/s, i.e. 7200 m/h, can be regarded as a lot 7.2 km in length whose transverse dimensions are less than a few tens of centimetres. While it is, strictly speaking, a three-dimensional object, two dimensions are negligible in comparison with the third, so the lot can be modelled as a simple function of time. This is equivalent to saying that the material can be represented by its projection onto a time axis. The elementary slice of material that has passed between times t and $t + dt$ is represented by its critical grade $a(t)$ or its contribution of heterogeneity of $h(t)$ (Chapter 8). To simplify matters, it has been assumed in Figure 11.1 that the flow rate is constant and consequently all the increments have the same mass. The profile of the curve

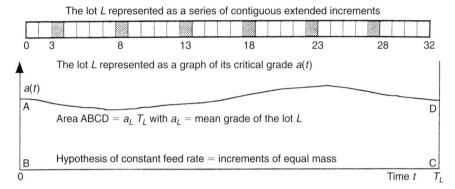

Figure 11.1   Two ways of representing the lot *L*

$a(t)$ shown in the lower part of the figure depends only on the time-scale. Measured over a short period its increase is imperceptible. Experimental evidence shows that if the fluctuations in the flow rate are no greater than $\pm 20\%$ of the mean flow, the curve $h(t)$ is in practice indistinguishable from that of $a(t)$.

Extracting an increment occupies a time interval Δt for its completion, the latter varying from a fraction of a second to a few seconds. This is represented by a few millimetres on the figure and shows clearly that the change in $a(t)$ in this interval is negligible. However, this does not imply that the model fails to account for the change. Within each increment, the latter being regarded as an independent entity, it appears as a component of the segregation and grouping error SGE and of the minimum practical error MPE. It will be shown in Section 12.9.1, which analyses the results of what are presented under the title of a 'variographic experiment', that these errors appear as a component of the intercept at the origin of the variogram. Thus, the models in zero and one dimension are linked together, without a break, at the scale of the increments.

11.3   CHARACTERISING THE HETEROGENEITY OF LOT *L*

The object of doing this is to connect the function $h(t)$ to the variance of the sampling error made by taking the sample E to be, for example, the assemblage of the increments numbers 3, 8, 13, 18, 23, 28 shown shaded in the upper half of Figure 11.1. If the mathematical expression for $h(t)$ were known from start to finish of the period during which lot L was in motion it would be a simple matter to do this, but it is never known. Thus the nature of the functions $h(t)$, which play such an important role in the one-dimensional model of sampling theory, needs to be studied.

Needless to say, moving streams of material are found and sampled in the majority of continuous process industries, particularly in those concerned with converting raw materials into finished products: marketable products whose quality needs to be controlled; solid wastes; liquid and multi-phase effluents that need to be controlled so as to protect the environment in compliance with national or international standards and local regulations. Sometimes attempts are made to improve a product's composition and so obtain a better price, which again calls for tight control of the process. The practical importance of these points is shown in this and the following chapters. The calculation of sampling variances of increments taken from moving streams is a problem that must be properly addressed, even though certain authors and standards committees resort to excessively simplistic solutions.

11.4 NATURE OF THE FUNCTION $x(t)$ [EITHER $a(t)$ OR $h(t)$]

The examination of many time series of results involving $x(t_q)$ has shown that they can never be expressed by simple algebraic expressions. A close study shows that they all have a double character:

1. They have a certain continuity that shows up only at a large scale, such as that shown by the profile of Figure 11.1 taken over its whole length
2. They also have a certain discontinuity which acts at the scale of the increments I_q and of their constituent elements, a discontinuity that simply reflects the essential discontinuity present in all materials and the random nature of analyses. It must be remembered that continuity is a mathematical concept that does not exist in the physical, material world.

To give a visual example, imagine a country scene of wooded hills of the kind often found in temperate climates. Seen from afar, the horizon looks like the profile of Figure 11.1. It reflects the continuity (more or less) of the shape of the earth on which the trees are growing. On the other hand, seen close to, the sawtooth silhouette of the trees and bushes is prominent, and their variable heights almost totally obscure the contours of the underlying bare earth. Taking this view, the earth is the element of continuity and the vegetation is the element of discontinuity.

When one also studies the broken line joining a series of points whose abscissa is a time-base and whose ordinate is a 'random number' of a kind that a table of random numbers or computer could generate, the analogy with the silhouette of the forest is striking. In the countryside described above, because of their form the trees introduce a discontinuous, quasi-random component. The 'quasi' aspect is emphasised. If the effectively random nature of the results of analyses are disregarded, then on the curves $x(t)$ as in the countryside example the discontinuous component has only the appearance and certain properties of randomness.

This leads to the recognition of three kinds of relationship between the functions $x(t)$ and the time variable t, as shown in Figure 11.2.

1. A continuous one (which can never be expressed mathematically). Whatever the value of t, the function $x_1(t)$ has a single, clearly defined value. This function is essentially continuous, i.e. continuous, with a finite number of discontinuities. Among the functions of this type there is one that plays a very particular role in sampling theory: a cyclic or quasi-periodic function (Figures 11.7 to 11.12).
2. A discontinuous or quasi-random one. Whatever the value of t, there is neither any connection nor any correlation between $x_2(t)$ and $x_2(t + \Delta t)$. The function $x_2(t)$ is discontinuous at every point and behaves as a random function.
3. A stochastic one, a hybrid resulting from the addition of the first two. For every value of t, there is a variable correlation between $x(t)$ and $x(t + \Delta t)$. But the function $x(t)$ remains discontinuous at every point.

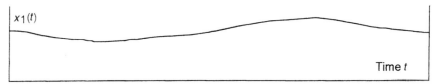

(a) Continuous process $x_1(t)$. The curve is continuous at every point. By analogy, $x_1(t)$ is a kind of carrier wave.

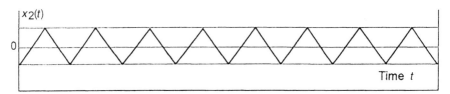

(b) Discontinuous or 'quasi-random' process $x_2(t)$. The curve is discontinuous at every point. The 'saw-teeth' are a symbolic representation of the amplitude of the 'quasi-random' process.

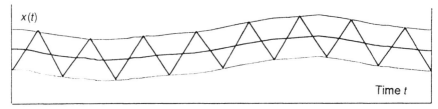

(c) Stochastic process: $x(t) = x_1(t) + x_2(t)$. The curve is discontinuous at every point. The saw-teeth are symbolic.

Figure 11.2   Continuous, discontinuous and stochastic processes

A philosophical digression: One cannot help thinking of J. Monod and his thesis *Chance and necessity*; of the random and of the continuous. Or of Darwin's theory of the evolution of the species. Or again of the English quotation 'God writes straight in crooked lines'. If these ideas are accepted, then life itself is a stochastic phenomenon, and everthing that goes with it, as well as many physical and other phenomena.

A more technical note: The adjective 'stochastic' is known to have at least two definitions and this may prove confusing. The first definition, given by some dictionaries and encyclopedias, is in fact perfectly synonymous with 'random'. If such were the case there would be no point in introducing a better expression than that. In 1960, M. S. Bartlett in his *Introduction to Stochastic Processes* (Cambridge University Press) wrote:

> By a stochastic process we shall in the first place mean some possible actual, e.g., physical process in the real world, that has some random or stochastic element involved in its structure.

Interpreting what seems to be a contradiction it can be seen that Bartlett felt the need to distinguish between that which is 'random' (i.e. 'purely random') and that which 'has (among other things) a random element'. To avoid any ambiguity the following definitions illustrated in Figure 11.2 (and see references [6] and [7] also) are proposed.

A process is said to be 'stochastic' if it is the result of superimposing two elements, one continuous, the other discontinuous; if it is represented by a function such as $x(t)$ it is the sum of two other functions:

- One essentially continuous, represented by $x_1(t)$ and
- One discontinuous, quasi-random represented by $x_2(t)$.

Experience has shown that the input from the two terms can be in any proportion, either of the two being the dominant one. In Figure 11.2 the continuous component is 'dominated' by the discontinuous, quasi-random one. Other examples will be given. This section began by 'synthesising' a stochastic function, but in some cases $x(t)$ can be analysed and split into its two components (see below at Section 12.14 and in Section 5.12 of reference [7]).

11.5 INTRODUCTION TO THE CHARACTERISATION OF THE FUNCTION $x(t)$

As the mathematical expression for the function $x(t)$ is never known, the best that can be done is to estimate it from a certain number of points. However, point values of $x(t)$ cannot be found exactly. Points are mathematical abstractions and one works on segments, usually taken at regular intervals such as the increments numbers 3, 8, ... 23, 28, etc. shown shaded in Figure 11.1. In practice these are (supposedly taken correctly) the constituent elements contained between two parallel planes perpendicular or oblique to the time axis (see Figure 6.6) and separated by Δt. Afterwards, their mean grade, their mass and their contribution to heterogeneity are estimated, and all subsequent work is based on these estimates.

The problem is how to characterise the fluctuations of the descriptive function $x'(t_q)$ (the prime in this notation indicates that it is not a true value—which is always unknown— but is instead an experimental estimate) that describe the qualitative and quantitative properties of the material that has passed between times $t_q - \Delta t/2$ and $t_q + \Delta t/2$.

Hardy spirits have no hesitation in confusing 'ordered series' with 'populations' and so are content to characterise the heterogeneity of a series by calculating the variance $s^2[x'(t_q)]$ from the values of $x'(t_q)$, as is so balefully recommended by the majority of sampling standards committees. This is an elementary fault whose practical consequences will be revealed in Section 12.13. The committees may be excused (?) to some extent by noting that the mathematical route leading from a set of estimates of the function $x'(t_q)$ to the variance of the sampling error comprises many not so evident intermediate steps, as will be demonstrated next.

11.6   CHARACTERISING THE FUNCTION $x(t)$: THE VARIOGRAM

11.6.1   Introduction

The variogram is the first function to be considered. It characterises the one-dimensional heterogeneity of material lying along an axis that may be geometric, in the manner proposed in 1960 by Georges Matheron [14] who was studying the sampling of orebodies, or time-based as will be proposed for characterising chronological series.

In contrast to a scalar variance, the variogram is a function of the distance between two points on some chosen axis that may be measured in units of length or time. This function expresses the correlation between any two points lying on the axis in terms of their distance apart or in some other way, i.e. the autocorrelation of the function.

Suppose that $Q = 60$ increments I_q have been taken at regular intervals T_0

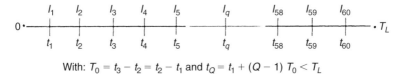

With: $T_0 = t_3 - t_2 = t_2 - t_1$ and $t_Q = t_1 + (Q - 1) T_0 < T_L$

Figure 11.3   Representation of a lot L by $Q = 60$ increments I_q

during the passage of the lot L between time $t = 0$ and time $t = T_L$. The increments I_q are weighed and analysed, and their heterogeneity $h_q = h(t_q)$ is calculated. (See Figure 11.3)

11.6.2 Definition of the variogram

For each value of $q + j < Q$: the increase $\Delta h(q + j, q)$ in the heterogeneity h_q between the final observation at time $t_{(q+j)}$ and the instant t_q at which the observations started, is calculated. By definition:

$$\Delta h(q + j, q) = h_{(q+j)} - h_q$$

These two points are separated on the time axis by an interval of $[t_{(q+j)} - t_q] = jT_0$, which is called the 'lag'. What is important is not the value or sign of each individual $\Delta h(q + j, q)$; it is instead their quadratic mean at each lag. This is why the function called the variogram $v(jT_0)$, or more simply $v(j)$, is defined as the semi-mean square of the increment:

$$v(j) = \frac{1}{2(Q - j)} \sum_q [\Delta h(q + j, q)]^2$$

Why use the semi-mean square and not the mean square? Suppose that the two values of h_q are taken at random from a population with zero mean and variance s^z. Each value of $v(j)$ as defined above would then be an estimator of the variance s^2. Thus, if the value of the mean square were used instead of the semi-mean square, it would be an estimator of $2s^2$.

11.6.3 Illustration of How to Calculate the Value of a Variogram at $j = 2$

The observations are set out in Figure 11.4. At lag $j = 2$ the lower time axis is displaced two T_0 intervals to the right. The values on the bottom line are subtracted from those immediately above them and the differences are squared.

$\Delta h\,(3, 1) = h_{(1+j)} - h_1 = h_3 - h_1$
$\Delta h\,(4, 2) = h_{(2+j)} - h_2 = h_4 - h_2$
$\Delta h\,(5, 3) = h_{(3+j)} - h_3 = h_5 - h_3$
$\Delta h\,(58, 56) = h_{(56+j)} - h_{56} = h_{58} - h_{56}$

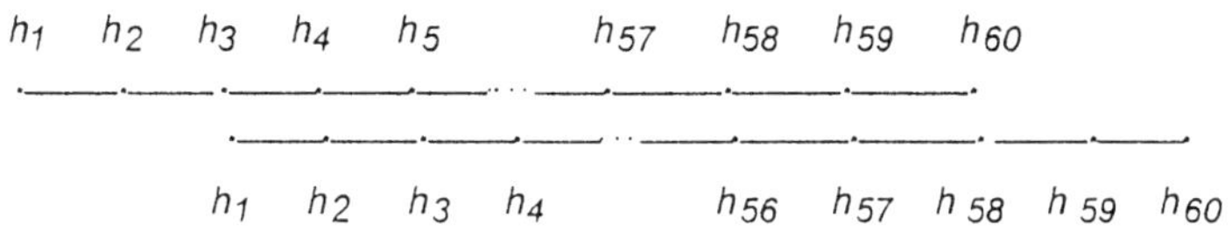

h_1 h_2 h_3 h_4 h_5 h_{57} h_{58} h_{59} h_{60}

        h_1 h_2 h_3 h_4 h_{56} h_{57} h_{58} h_{59} h_{60}

Figure 11.4   Calculation of the variogram at lag $j = 2$

$$\Delta h\,(59,\,57) = h_{(57+j)} - h_{57} = h_{59} - h_{57}$$
$$\Delta h\,(60,\,58) = h_{(58+j)} - h_{58} = h_{60} - h_{58}$$

The number of differences $Q - j = 60 - 2 = 58$. This number is the denominator of the mean square. The above calculations show how variograms were calculated 'by hand' in 1960. With a computer, even a modest one, the saving in time is enormous.

11.7   EXAMPLES OF VARIOGRAMS

11.7.1   Variogram Increasing Linearly

Figure 11.5 shows the variogram of the feed to a uranium ore treatment plant. $Q = 60$ increments were taken at intervals of 2 mn and the variogram of h_q was calculated for values of j from 1 to 30. The variogram shows regular behaviour. It increases quasi-linearly for values of j from 1 to 19, and a quasi-random term is present at all lags.

The variance of the population of h_q is also calculated. This variance is the model of a variogram described by a horizontal line defined by $v(j) = $ constant $= s^2(h_q)$ that would be seen if, instead of being autocorrelated, the values of h_q were taken at random from a population with a mean of zero and a variance $s^2(h_q)$. This line is called the 'sill' of the variogram. It characterises the global heterogeneity of the series. The variogram, on the other hand, characterises the sequential heterogeneity of the series.

Up to $j = 18$, i.e. up to $jT_0 = 36$ mn, the variogram lies below the sill and

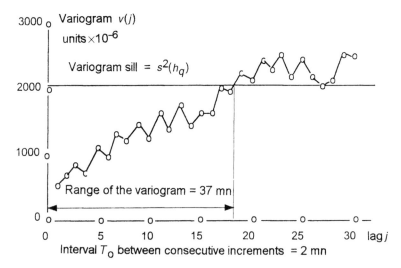

Figure 11.5   Variogram of the feed to a uranium ore mill

increases steadily. After that it oscillates at random about or near the sill. The abscissa of the point of intersection of the variogram and the sill is called the 'range' of the variogram (here 37 mn). It is the time interval beyond which no correlation between the observations can be found. The last point ($j = 30$) has been calculated with $Q - j = 30$ degrees of freedom. Beyond that value the significance of the points decreases progressively. It can also be seen that, by definition, the variogram does not give an estimate of $v(j)$ at $j = 0$. This is an important point to which more attention will be paid later, because the estimation of sampling variances requires a knowledge of the value of $v(0)$, the variogram's intercept on the ordinate.

11.7.2 Variogram Increasing more Irregularly

Figure 11.6 is the variogram of the feed to cement kiln. The critical element in this case was the grade a_q of CaO. $Q = 60$ increments were taken at intervals of 2 mn. They were weighed and their contributions h_q to the heterogeneity of the CaO grade calculated. Because of the small variations in the feed rate, the variogram of the grade a_q is practically the same as that of h_q. The variogram shows complex behaviour.

11.7.3 A Simple Periodic Variogram

Figure 11.7 shows the variogram of the masses of the increments taken in the previous section. From note 2 of Section 8.2.1, the contribution of heterogeneity h_q'' relative to the masses can be defined as:

$$h_q'' = (M_q - M_{q*})/M_{q*}$$

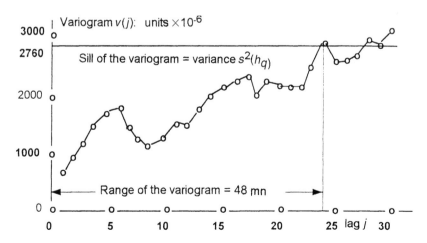

Figure 11.6 Variogram of the feed to a cement kiln. Heterogeneity h_q in terms of the CaO grade. Sampling interval (lag) $T_0 = 2$ mn

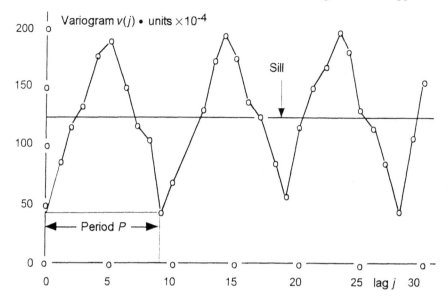

Figure 11.7   Cement kiln feedstuff. Variogram of the masses of the increments

The variogram's markedly cyclic nature is not surprising. In fact, the feed rate varied between a maximum and a minimum of (variable) period P covering nine or ten 2 minute intervals. Theory shows that the variogram of $y = \sin x$ of period P is itself sinusoidal and has the same period P. The first maximum (an estimate of $P/2$) is at $j = 5$; the first minimum (an estimate of P) is at $j = 9$; a second maximum (an estimate of $3P/2$) is at $j = 14$; a second minimum (an estimate of $2P$) is at $j = 19$; a third maximum (an estimate of $5P/2$) is at $j = 23$; and finally a third minimum (an estimate of $3P$) is at $j = 29$. This set of results gives six estimates of P between 18 and 20 mn (Figure 11.7).

Three things can be noted:

1. The masses of the increments showed no obvious pattern of cyclic behaviour; this was revealed by the variogram. This is due to the fact that large numbers of degrees of freedom were used (between 59 and 30 in this experiment) so the considerable random component that tended to hide the cyclic pattern in the feed rate, and consequently in the masses of the increments, was significantly attenuated by the variogram.
2. The sill no longer acts as an asymptote.
3. When taking samples it is of the utmost importance to detect the presence of cyclic fluctuations in quality and/or feed rate, because if samples are taken at the same time interval $T_0 = P$ on a sine wave, the increments are all taken at the same point on the cycle and the sampling variance is calculated wrongly.

This situation must be avoided at all costs. Thus, the variogram is shown to be an excellent tool for analysing and discovering the presence of one or several cyclic components in stochastic phenomena.

11.7.4 A Complex Periodic Variogram

The variogram in Figure 11.8 was one of the first to be constructed in 1962 from the feed to a zinc ore mill. It is based on increments taken at intervals of 20 mn and shows a cyclic component with a period of about 10 lags or 200 mn, which corresponds neither to any natural cyclic behaviour in the run of mine ore nor to the crushing and grinding section of the mill. The mine was shut down shortly after this experiment so the physical cause of these clearly marked fluctuations remains unexplained.

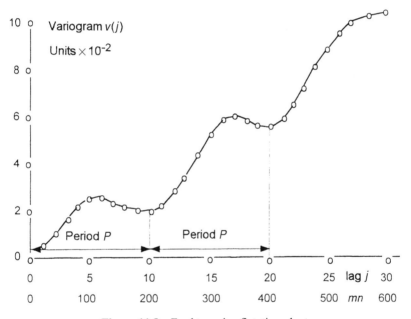

Figure 11.8   Feed to a zinc flotation plant

11.7.5 Other Examples of Cyclic Variograms

A theory of one-dimensional homogenisation is presented in Chapter 35 of references [6] and [7], better known in industry under the name of 'pre-homogenisation' or 'bed-blending'.

In 1978, six months after the development of this theory, the author had an opportunity to organise a full-scale experiment to (successfully) check its validity. On this occasion the frequency of occurrence of cyclic variograms was

confirmed. The most typical are shown here. Variograms were calculated of the
raw materials fed to a cement kiln, of the grades in SiO_2, CaO, Fe_2O_3 and
Al_2O_3.

The material travelled on a 26 m belt conveyor, placed just upstream from the
blending bed, that was moving at 1 m/s. The conveyor was stopped and the
material on it was divided into 26 segments each 1 metre in length and thus
representing one second of feed. Each segment was weighed, reduced and
analysed. The object at this stage was simply to calculate their variances. No one
suspected the existence of cyclic fluctuations with a period of about 4 seconds,
which Figures 11.9 to 11.12 show so clearly (they are not significant in practical
terms). Each of the variograms on its own is interesting, but the similarity
between them is even more so.

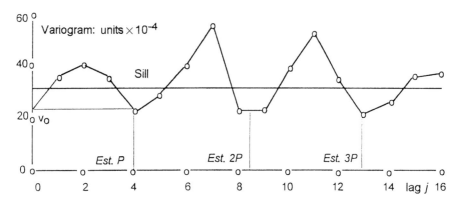

Figure 11.9   Cement kiln feed. SiO_2 grade $^*T_0 = 1$ second

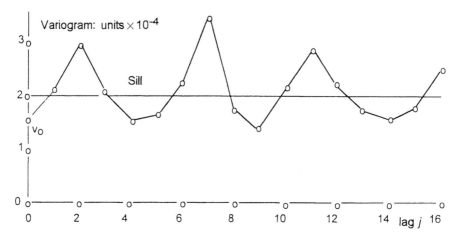

Figure 11.10   Cement kiln feed. CaO grade $^*T_0 = 1$ second

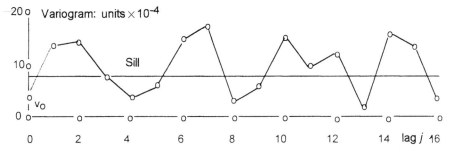

Figure 11.11 Cement kiln feed. Fe_2O_3 grade $^*T_0 = 1$ second

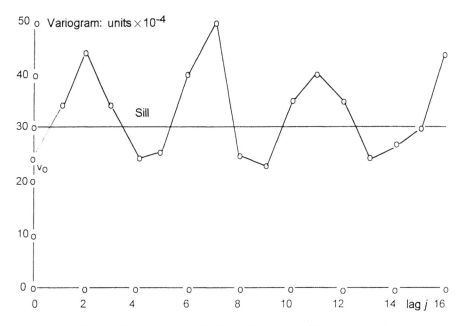

Figure 11.12 Cement kiln feed. Al_2O_3 grade $^*T_0 = 1$ second

The most remarkable feature of the four variograms is their similiarity of form. For example, they all have the same first minimum point, which gives an estimate of the period, lying between 4 and 5 seconds; a second minimum between 8 and 9 seconds; and a third between 13 and 14 seconds, all of which confirm a common period of about 4.5 seconds. Their four maxima behave likewise. Their similarity is even more remarkable in view of the fairly low number of degrees of freedom; 25 at $j = 1$ and only 10 at $j = 16$. While it was possible to forecast a similar cycle for the fluctuations in the feed rate because of the way in which the material was fed onto the belt, it was completely impossible to foresee the very close correlation between the grades of the four principal components.

11.8   THE VARIOGRAM AS A TOOL FOR THE ANALYSIS OF HETEROGENEITY

These examples lead naturally to inquiries into the possibility of finding wider uses for the variogram. It has been considered so far to be a necessary step towards the calculation of sampling variances. Experience has shown that it is also a powerful tool for the analysis of the one-dimensional heterogeneity of a time series of data. In particular, it has allowed cyclic fluctations to be detected that would otherwise have remained undetected by other means, and which may be of considerable importance in the optimisation of certain continuous process operations.

Moreover, it has played a decisive role in the development of one-dimensional blending processes. It is thus a very powerful, albeit neglected, tool.

11.9   DESIGNING A VARIOGRAPHIC EXPERIMENT

A 'variographic experiment' is one designed to construct one or several variograms. When planning the experiment it is essential to first define its object precisely. It could, for example, be concerned with the design of a sampling installation, and this would require two series of increments to be taken:

- A first series of 60 to 100 increments taken at a constant interval of between 30 seconds and 2 minutes, designed to span (as will be seen in Chapter 12) the sampling intervals most commonly used in sampling practice, which are rarely greater than 20 mn.
- A second series of 30 to 50 increments taken as closely together as possible (1 to 5 seconds, or as closely as local conditions allow) to make an accurate estimate of the ordinate $v(0)$ at the origin and to determine $v(\varepsilon)$ over $\varepsilon = 1$ to 5 seconds. When cyclic fluctuations of the kind described in Section 11.7.5 above are absent it is enough to calculate the variance of the data so collected. However, experience has shown that it is always instructive to determine the variogram as this gives an estimate of $v(0)$ which is nearer to reality than the variance of the population, which corresponds to the sill of the variogram.

It can also be used to examine the longitudinal heterogeneity of a lot of moving material when checking how well a 'bed-blending' installation is working (Chapters 35 of references [6] and [7]). In these, variograms were determined based on intervals of 90 and even 120 mn. In every case it is imperative to define clearly the object of a variographic experiment.

The author is wholly convinced that the variogram is an extremely powerful analytical tool whose utility has yet to be recognised by the world of industry for the purpose of optimising the running of its production and processing plants; of minimising working costs and maximising added value.

Chapter 12

Sampling: The One-dimensional Model

12.1 INTRODUCTION

It has been shown how the one-dimensional heterogeneity of chronologically ordered sets or time series can be characterised with the aid of a very powerful tool: the variogram. Next, the variogram function will be used to develop other functions that are needed for the estimation of sampling variances. For this purpose the following have to be formulated:

- The auxiliary functions of the variogram: its single and double integrals
- The three most common ways in which increments are taken in terms of the time intervals between them
- The error-generating functions, and from these
- The integration variance associated with each of the three ways of taking the increments.

12.2 THE VARIOGRAM'S AUXILIARY FUNCTIONS

The process is a fairly long one, beginning with the series of observations $v(j)$ of the variogram and ending with the integration variance. Four auxiliary functions have to be defined:

- The single integral $S(j)$ of the variogram
- The mean value $w(j)$ of $S(j)$
- The double integral $S'(j)$ of the variogram
- The mean value $w'(j)$ of $S'(j)$

12.3   THE MEAN VALUE $w(j)$ OF THE SINGLE INTEGRAL OF THE VARIOGRAM

By definition (and see Appendix 3 for an explanation of the nomenclature):

$$w(j) = \frac{1}{j} S(j) = \frac{1}{j} \int_0^j v(j') \, dj'$$

12.4   THE MEAN $w'(j)$ OF THE DOUBLE INTEGRAL OF THE VARIOGRAM

This is denoted by $w'(j)$. By definition (see Appendix 3):

$$w'(j) = \frac{1}{j} S'(j) = \frac{1}{j} \int_0^j w(j') \, dj' = \frac{2}{j^2} \int_0^j dj' \int_0^{j'} v(j'') \, dj''$$

12.5   THE USUAL INTERVALS AT WHICH INCREMENTS ARE TAKEN

At this point on the journey to the estimation of sampling variances there need to be considered the times at which the increments can be selected along the curve $x(t)$. Omitting the case, unhappily an all too common one, where selection is left to the arbitrary whim of the operator (note: arbitrary is not random) and increments are taken 'from time to time', there is an infinite number of 'organised' selection modes, among which three are considered to be the most usual ways of selecting points along a time axis, or, in other words, ways of selecting the instants at which increments are taken.

- *Systematic selection* (denoted by sy) which can best be defined as the selection of points spaced at a uniform interval T_{sy}, the time of the first increment having been selected at random in the interval between $t = 0$ and $t = T_{sy}$. It is by far the commonest selection mode used in industry. The constant time interval is guaranteed by a simple timer and the random choice of the timing of the first increment arises from the absence of correlation between its timing and the properties of the material flowing at that instant.
- *Stratified random selection* (denoted by st). The total time $D_L = [t = 0, t = T_L]$ during which lot L is flowing is divided into a number of intervals of uniform duration T_{st}. A point is selected at random within each interval to represent the material that has passed the sampling point during that interval. The object of this type of sampling is to avoid the serious risk of discovering later that cyclic fluctuations are present and a uniform interval T_{sy} of systematic sampling had been used that is an integer multiple of the period

P of the phenomenon. This technique has been described since the 1960s and has been used in cement works where such fluctuations have been observed, after they had been detected due to their variograms. When designing a processing plant, if cyclic fluctuations are suspected in the feed to it, stratified random sampling is the safe solution.

- *Completely random sampling* (denoted by ha). First, the number of increments Q_{ha} to take from lot L is decided. Then, with the aid of a book of random number tables or a random number generator, the Q_{ha} instants at which the increments are taken from within the time D_L are determined. This sampling mode is never used, unless a comparison needs to be made between the nonsensical results of applying the statistics of non-ordered populations to a chronologically ordered series of observations.

12.6 THE ERROR-GENERATING FUNCTIONS $W(j)$

In the development of expressions for the integration variances, Chapters 13 of references [6] and [7] introduced what are called error-generating functions, $W(j)$, derived from the auxiliary functions $w(j)$ and $w'(j)$ of the variogram, and which are characteristic of the three ways of selecting increments as described in the previous section. They play, in chronologically ordered series, the same role as the variance V of a population. For a sample made up of Q increments, the integration variance $s^2(\text{IE})$ can be expressed in the case of a series as $W(j)/Q$, and for a population as V/Q. There are thus three error-generating functions:

- Systematic sampling: $\quad W_{sy}(j) = 2w(j/2) - w'(j)$
- Stratified random sampling: $\quad W_{st}(j) = w'(j)$
- Completely random sampling: $\quad W_{ha} = s^2(h_q) = DH_L = \text{const.}$

This last constant is none other than the sill of the variogram.

12.7 THE VARIANCE $s^2(\text{IE})$ OF THE INTEGRATION ERROR IE

For each of the three increment selection modes, if T_{sy} is the systematic interval, T_{st} the length of the interval and Q_{ha} the number of increments taken at random, the integration variance $s^2(\text{IE})$ can be written, with

$$Q = T_L/T_{sy} \text{ or } Q = T_L/T_{st}:$$

- Systematic: $\quad s^2(\text{IE})_{sy} = W_{sy}(T_{sy})/Q = W_{sy}(T_{sy})T_{sy}/T_L$
- Stratified random: $\quad s^2(\text{IE})_{st} = W_{st}(T_{st})/Q = W_{st}(T_{st})T_{st}/T_L$
- Completely random: $\quad s^2(\text{IE})_{ha} = W_{ha}/Q_{ha} = s^2(h_q)/Q_{ha}$

The integration variances so calculated are slightly overestimated because the experimental data contain the errors of sample preparation and analysis (these are relatively small and vary with j).

12.8 PRACTICAL CALCULATION OF THE AUXILIARY FUNCTIONS

The problem now is how to calculate the auxiliary functions $w(j)$ and $w'(j)$, on the one hand, and the error generating functions $W(j)$, on the other, given a set of points on a variogram: $v(1)$, $v(2)$, ... $v(Q - Q_0)$. Historically, two methods have been proposed:

- To model the variogram by fitting some algebraic expression: this is the method used by geostatisticians and was the method used by the author up to 1982. This method is too imprecise because of its artificiality. It was abandoned in favour of:
- The point-by-point calculation of the auxiliary functions and error-generating functions: this is the method used since 1982 and published for the first time in reference [6] in 1988. Obviously, it can be easily computerised and from now on it is unquestionably the recommended method.

Both methods require an estimate of the ordinate $v(0)$ at the origin which, because of its algebraic formulation, the variogram is unable to give. Thus, the next section will show how this is to be done.

12.9 ESTIMATION OF THE ORDINATE $v(0)$ OF THE VARIOGRAM (OR INTERCEPT)

Several methods of estimating $v(0)$ can be thought of but none of them is strictly objective. It is first necessary to investigate the physical significance and the content of this ordinate.

12.9.1 Significance and Content of the Intercept

Geostatisticians have given $v(0)$ the colourful name of the 'nugget effect'. This severely limits, even in geostatistics, a full appreciation of what is represented by this parameter, which is much more than its name suggests. In fact the ordinate $v(0)$, which is always positive, is the sum of several components, in particular:

- The variance of the fundamental error FE generated during the extraction of the increments at the time of the variographic experiment
- The variance of the segregation and grouping error SGE generated at the same time

- The variance of all the other components of the sampling error
- The variance of the different sampling errors generated during the stages of sample reduction from the primary increments (maybe of several tens of kilograms each) to the final selection of the material taken for analysis
- The variance of the true analytical error AE.

Of all these errors, the only one that the expression 'nugget effect' calls to mind is SGE. Indeed, in something that seems to be a kind of rag-bag of errors, all the errors related to the discrete structures of materials can be found. The only component outside the rag-bag is the analytical error AE. Without in any way being critical of our geostatistical colleagues, they behave as though they are dealing with a series of actual grades—always unknown—whereas like us they are in fact dealing with random estimates introduced by selection, preparation (crushing, grinding, drying, etc.) and of analysis. All the errors so generated are to be found in the intercept $v(0)$ of the variogram.

It may be noted that the intercept $v(0)$, through the fundamental error FE and the segregation and grouping error SGE related to the extraction of increments during a variographic experiment, is the point where the one-dimensional model of the integration error IE (somewhat abstract because of its algebraic formulation) links up with the zero-dimensional model, in the development of which the errors FE and SGE were defined in Part II.

12.9.2 Simple Extrapolation of the Points $v(3)$, $v(2)$, $v(1)$ to Zero

When the first points of the variogram show a regular behaviour like those in Figures 11.5 and 11.6, no serious errors are incurred by extrapolating them to zero. In both cases excellent estimates are made with (coincidentally) values of 600.10^{-6}. As will be seen, a more precise estimate than this is not necessary.

12.9.3 Calculation of the Variogram $v(j)$ with j Extending over Several Seconds

If this problem is to be solved more precisely it is recommended that a set of at least 30 increments is taken at very short intervals of one to five seconds or as closely as local conditions allow. The reader is referred again to the four variograms of the feed to a cement kiln (Figures 11.9 to 11.12). Experience has shown that a stopped belt, neither too long (26 m) nor too short, is ideal for this purpose. The 26 m were split into 26 segments each one metre in length which, as the belt was travelling at 1 m/s, had their centres spaced one second apart and whose masses corresponded to those taken by an industrial sampler (about 100 kg).

Indeed, it makes no sense to compare the variograms or variances of series or of populations of samples whose masses have different orders of magnitude. This is obvious when one thinks of the components of $v(0)$ and of the importance of the mass of a sample (in terms of a single increment) in the calculation of the fundamental variance.

The series of closely spaced increments was taken with the object of estimating the value of $v(1$ second) and from this to assimilate $v(\varepsilon)$ to $(v(0)$. The discovery of the pattern of cyclic behaviour of period about 4.5 seconds, shown in Section 11.7.5, was unexpected. Naturally, this led to the analysis of the ordinate $v(0)$ at the origin in those cases where the variograms showed a simple cyclic behaviour, such as those illustrated in Figures 11.7 and in 11.9 to 11.12.

A theoretical study of the variogram of periodic functions of period P that were more or less sinusoidal (maxima and minima alternating regularly—see Sections 5.9.3 and 5.9.4 of reference [7])—led to the conclusion that the variogram of all these functions showed a distinctly cyclic character of the same period P.

It is easy to show that, for a sinusoidal function, the variogram at $v(0)$ is equal to zero, as it is at $v(P)$, $v(2P)$, etc. Knowing the value of $v'(P)$, the experimental value with the highest number of degrees of freedom and therefore the point whose estimated value is statisically the most reliable, it is recommended that this value is used as a reliable estimate $v'(0)$ of $v(0)$. It can be seen in Figure 11.7 that the value of the first minimum on the variogram gives an estimate of $v'(0) = v(P) = 45.10^{-4}$, a value hardly different from that which would have been obtained by extrapolation (about 50.10^{-4}).

On the other hand, in Figures 11.8 to 11.10 the value $v'(0) = v(P)$ is much more reliable than would have been obtained by extrapolation. Thus, for SiO_2: $v'(0) = 22.10^{-4}$; for CaO: $v'(0) = 1.6.10^{-4}$; for Fe_2O_3: $v'(0) = 5.10^{-4}$ and for Al_2O_3: $v'(0) = 24.10^{-4}$.

As for using a variance numerically equal to the sill, this solution is clearly unsatisfactory when the cyclic character of the variogram is confirmed. In contrast, when the variogram looks 'flat' (Section 5.9.1 of reference [7]) and its points are scattered apparently at random above and below the sill, this solution, which uses the highest number of degrees of freedom $(Q - 1)$, gives the most reliable estimate of $v(0)$.

12.9.4   Duplicating the Increments

This is a very specific and aesthetically satisfying method which takes duplicate increments at intervals of 30 seconds to 2 minutes. Provided enough material is available it is easy to:

• Divide each increment into two by splitting (the increments should be double their normal mass) or
• Use two identical increment cutters or
• Repeat the operation by taking increments at intervals of a few seconds

The three methods, all of which have been used in practice, are equally satisfactory. The last is the simplest to use. All three methods give Q pairs of matching increments and Q differences between the estimated or calculated values of a_q or h_q. The variance of this population of differences is an excellent estimator of $2v(0)$.

To conclude, these are the possible methods and the local constraints that serve as guides to the best way of determining the value of $v(0)$.

12.10 POINT-BY-POINT CALCULATION OF AN INTEGRAL

No time will be wasted on demonstrating the superiority of this method over the one that relies on the algebraic modelling of $v(0)$ and the algebraic calculation of its single and double integrals. An interested reader can find it in Sections 5.5 and following in reference [7].

The technique on which the estimation of the auxiliary functions $w(j)$ and $w'(j)$ depend is based entirely on the point-by-point estimation of an integral. The technique is explained by Figure 12.1. It rests on two hypotheses that appear to be the most realistic and the simplest of any:

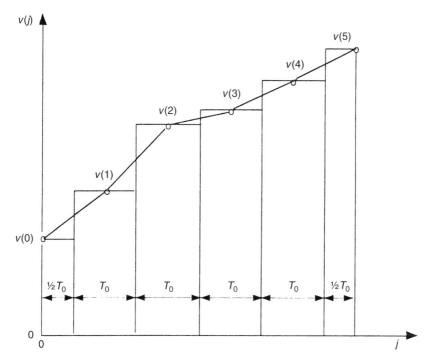

Figure 12.1 Practical estimation of a variogram's integral

1. The variogram passes through all the experimental points $v(j)$, and through the estimated value of $v(0)$.
2. The variogram consists of straight lines connecting its points.

In keeping with these two hypotheses the integral of $v(j)$ between $j = 0$ and $j = 5$ is equal to the area enclosed by the lines joining the points $v(j)$, and the abscissae axes. Some elementary geometrical observations show that this area is strictly equal to the area enclosed by the step-wise line joining the points $v(0)$ to $v(5)$ and the abscissae axes.

This last area $S(5)$ can be written:

$$S(5) = \tfrac{1}{2}T_0 v(0) + T_0 v(1) + T_0 v(2) + T_0 v(3) + T_0 v(4) + \tfrac{1}{2}T_0 v(5)$$

from which it follows that:

$$w(5) = \tfrac{1}{5}\int_0^5 v(j)\,dj \sim \tfrac{1}{5}S(5) \quad \text{with} \quad w(0) = v(0)$$

Substituting $w(j)$ into $v(j)$ gives the area S' and the estimate of $w'(j)$:

$$w'(5) = \tfrac{1}{5}\int_0^5 w(j)\,dj \sim \tfrac{1}{5}S'(5) \quad \text{with} \quad w'(0) = w(0) = v(0)$$

This technique is extremely easy to use and to program directly after calculating the values of $v(j)$.

12.11 POINT-BY-POINT CALCULATION OF THE AUXILIARY FUNCTIONS AND ERROR-GENERATING FUNCTIONS (EGF)

The variogram is of the feed to a uranium treatment plant.

$$\text{Lag}\,T_0 = 2 \text{ mn} * \text{ Units} \times 10^{-6} * \text{ Sill } s^2(h_q) = 1781.10^{-6}$$

For non-integer values of j, which are needed to calculate $w(j/2)$ and $W_{sy}(j)$, the values of $w(j)$ appearing in Table 12.1 have been calculated by linear interpolation.

In practice, the first thing to be done is to estimate $v(0)$, for example by extrapolating the values of $v(4)$, $v(3)$, $v(2)$ and $v(1)$, unless there is an independently determined value of $v(\varepsilon)$ that can be considered to be a reliable estimate of $v(0)$. The next step is to proceed by iteration, as follows. The following values are input and initialised thus:

$$w'(0) = w(0) = v(0) \text{ so that}$$
$$S'(0) = S(0) = 0$$

and then calculate, for $j \geqslant 1$:

Table 12.1

Lag	Variogram		Variogram auxiliary functions				Error-generating functions		
j	$v(j)$	$S(j)$	$w(j)$	$S'(j)$	$w'(j)$	$2w(j/2)$	$W_{sy}(j)$	$W_{st}(j)$	W_{ha}
0	600	0	600	0	600	1200	Indeterminate		
0.5	628	307	614						
1	655	628	628	314	628	1228	600	628	1781
1.5	719	971	647						
2	782	1346	673	1301	651	1256	605	651	1781
2.5	843	1752	701						
3	904	2189	730	3069	682	1296	614	682	1781
3.5	853	2628	751						
4	802	3042	761	5685	710	1346	636	710	1781
4.5	924	3473	772						
5	1045	3966	793	9189	735	1402	667	735	1781
5.5	1030	4485	815						
6	1014	4996	833	13 670	759	1460	701	759	1781
6.5	1131	5481	843						
7	1248	6127	875	19 232	789	1502	717	789	1781
7.5	1249	6750	900						
8	1249	7376	922	25 984	812	1522	710	812	1781

$$S(j) = S(j-1) + \tfrac{1}{2}v(j-1) + \tfrac{1}{2}v(j)$$

$$w(j) = S(j)/j$$

$$S'(j) = S'(j-1) + \tfrac{1}{2}S(j-1) + \tfrac{1}{2}S(j)$$

$$w'(j) = 2S'(j)/j^2$$

The value of $2w(j/2)$ is also needed to calculate $W_{sy}(j)$.
If j_0 is an integer, then:

$$\text{if } j \text{ is even: } j = 2j_0\text{: } 2w(j/2) = 2w(j_0)$$

$$\text{or } j \text{ is odd: } j = 2j_0 + 1\text{: } S(j_0 + \tfrac{1}{2}) = S(j_0) + \tfrac{1}{4}v(j_0) + \tfrac{1}{4}v(j_0 + \tfrac{1}{2})$$

$$2w(j/2) = 2S(j_0 + \tfrac{1}{2})/(j_0 + \tfrac{1}{2})$$

The values of $v(j_0 + \tfrac{1}{2})$ and $S(j_0 + \tfrac{1}{2})$ are calculated by linear interpolation

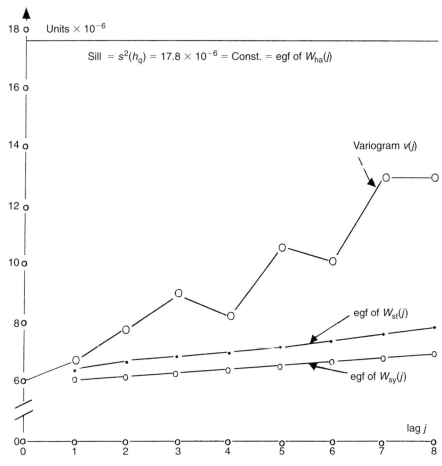

Figure 12.2   The variogram and its error-generating functions (egf)

12.12   THE VARIOGRAM AS A DETECTOR OF AUTOCORRELATION

Autocorrelation is an extremely important factor when sampling a stochastic function. Estimates of integration variances can be two or three times too high if autocorrelation is neglected and chronologically ordered data are treated as though they were simple populations characterised only by their variances, as is so shamelessly done by the majority of sampling standards. The following simulation is offered in evidence to illustrate this point.

- An actual variogram $v(j)$ is used, that of Figure 11.6.
- The values of h_m were redistributed randomly and given a new index m',

thereby destroying the series' autocorrelation and making it a random one.
- The variogram $v'(j)$ of the non-autocorrelated values $h_{m'}$ was calculated so as to compare it with the autocorrelated series.
- The variograms of $v(j)$ and $v'(j)$, together with their corresponding error-generating functions $W_{sy}(j)$ and W_{ha} (their common sill), are shown in Figure 12.3.

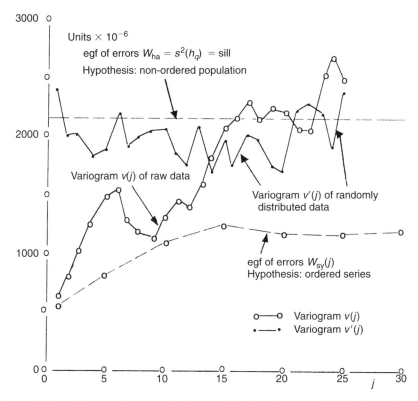

Figure 12.3 Influence of autocorrelation on ordered series as revealed by the variogram, and the corresponding egf

12.13 CONSEQUENCES OF THE ERROR MADE BY TREATING AN ORDERED SERIES AS A NON-ORDERED POPULATION

This faulty logic, common in sampling standards, causes a serious overestimation of the variance of the integration error. This variance is proportional to the error-generating functions (egf) shown by broken lines in Figure 12.3. In this example, which is typical, it can be seen that the error-generating function of random sampling, which, by definition, is none other than the common sill of the two variograms, is constant and equal to 2144.10^{-6}, whereas that from systematic sampling rises from 550 to 1200.10^{-6}, i.e. it is 3.9 to 1.8 times less.

Figure 12.4 shows the curves corresponding to the 95% confidence limits (assuming a normal distribution of errors) as functions of the number Q of increments taken over a period of 2 hours, for the three methods of selection.

The figure shows that, based on the variogram of Figure 12.3, to obtain, for example, a relative 95% confidence interval of $\pm 1\%$ (assuming a normal distribution), the following numbers of increments need to be taken over a two-hour period:

- sy mode: 25 increments at a uniform interval of 4.8 mn
- st mode: 28 increments at mean intervals of 4.3 mn
- ha mode: 71 increments at mean intervals of 1.7 mn.

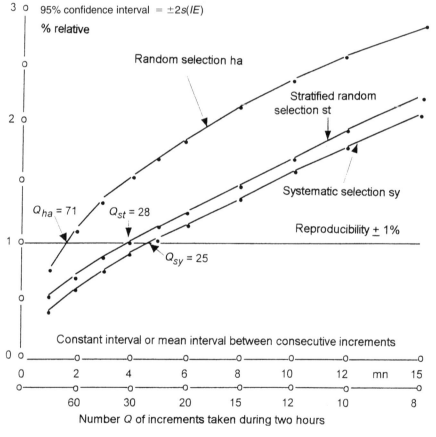

Figure 12.4 Number of increments needed to achieve a given reproductibility, according to the selection method used

If an ISO standard had been followed that relies on the variance of a population of grades which in reality belong to an ordered series of the kind studied here, this variance is equal to 1781.10^{-6}, so if a value of 0.5% or 5.10^{-3} had been chosen as the target relative standard deviation, the admissible variance is thus 25.10^{-6}, and this calls for 71 increments to be taken. If the variogram and its auxiliary functions had been calculated, the easiest and quickest of any task for even the most modest of personal computers, it can be seen that 25 systematic increments would have been enough to achieve the same reproducibility . If the possibility of cyclic fluctuations had been suspected it would have been easy enough to change to stratified random sampling by cutting the time during which the lot is flowing, say 120 mn, into 28 periods of 4.3 minutes each (or, perhaps, into 30 periods of 4 minutes each) and taking an increment at random within each 4-minute period.

The prudent reader will point out that it is always useful to use a safety factor, even one as large as 2.84, but economical readers will reply that if the increments weigh 100 kg on average, it is one thing to treat 1.25 tonnes of sample per hour and quite another to treat 3.55 tonnes per hour. They would soon make up their minds on that point and choose a solution that is both economical and more satisfying scientifically: systematic sampling at intervals of 4.8 minutes.

The 'safety factor' of 2.84 is a number whose magnitude (between 2 and 3) has appeared repeatedly in the author's experience. It can be regarded as a kind of 'coefficient of inadequacy' in those standards that insist on confusing a population with a chronologically ordered series.

12.14 DECOMPOSITION OF A STOCHASTIC SERIES INTO ITS TWO COMPONENTS

It was seen in Section 11.4 that stochastic data, whose contributions h_m to heterogeneity have been studied using variograms, can be considered as the sum of two components: an essentially continuous one designated by h_{m1} and another that is discontinuous at every point, quasi-random, and designated by h_{m2}. A striking example illustrating this is given in Section 5.12 of reference [7] and the reader is referred to it in order to avoid overburdening this book, which is intended to be an abridged version. Decomposition is based on the properties of the moving average.

Everyone knows that the variances of random data can be reduced by increasing their numbers. To illustrate this point, consider the series of 60 values of h_m whose variogram has been studied above and whose moving average has been calculated on the basis of 11 consecutive values. It is convenient to use an odd number for the amplitude or range R of the moving average, so that the means are allocated to the median positions within the range, in this case the sixth. Call h_m^* this moving average. The mean of the first eleven values of h_m^* can be written as h_6^*. Moving averages can only be calculated for values between $m = 6$ and

$m = 54$. For the five first and the five last values of the series moving averages over smaller ranges have been determined, with a small reduction in their precision. Thus: $R = 9$ for $m = 5$; $R = 7$ for $m = 4$; $R = 5$ for $m = 3$; $R = 3$ for $m = 2$; and finally $R = 1$ for $m = 1$.

The mean h_m^* is an estimate of the continuous component h_{m1} and the difference $(h_m - h_{m1})$ is an estimate of the discontinuous component h_{m2}.

12.15   VARIOGRAPHY AND THE CONTROL OF TREATMENT PLANTS

The expression 'treatment plants' includes all those industrial units that receive raw materials and treat them by continuous processes, changing them into different products, some valuable, some waste, and others being recycled at some point in the process. It is customary in such plants to draw up a 'balance sheet' of the process, so as to control carefully the quality and quantity of products entering and leaving the plant. Quality control is most often done by analysing increments taken at regular intervals, frequently by doubtful methods, but it will be assumed for the time being that the increments are both probabilistic and correct (see Chapter 4). With the development of modern analytical methods it is not unusual to find rapid 'on-line' analysers installed which (and assuming they give accurate and reliable results, an often optimistic hypothesis), although expensive to buy, once installed take frequent measurements at marginal cost and so give impressive series of chronologically ordered data.

They are often linked to computers as well. Loading an algorithm into a computer to calculate variograms of freshly gathered data is no problem. It allows continuous control of the sampling variance of each product entering, leaving or being recycled, and it checks on their mutual accordance. If the data and their confidence limits do not agree with each other, in other words if they bring into question (as has been the case on many occasions) the material balance in the plant, it is the responsibility of the person in charge to examine the reliability of the means by which the plant is being controlled and to remedy any defects. Computer print-outs give permanent records that can be studied at leisure. If the data so gathered are compatible and no malfunctions have been discovered in the control process then, thanks to the analytical power of the variogram, valuable information on each of the products and on how well the plant is running is given to the producers at little cost, and on the means by which the process may possibly be improved. Details of how this has been applied to bed-blending are given in Chapters 35 of references [6] and [7]. It is a promising area that remains to be developed further.

Part IV

THE QUANTITATIVE APPROACH

OTHER APPLICATIONS OF SAMPLING THEORY

Chapter 13

Measurement of Mass by Proportional Sampling

Setting to one side the notion of sampling in its strict sense the author is convinced that, among all the practical applications of the generalised theory of sampling, what is presented next is of the greatest importance to industry because dynamic weighing as practised at present is totally lacking in reliability.

13.1 JUSTIFICATION FOR WEIGHING BY PROPORTIONAL SAMPLING

Studies of proportional sampling first began in 1955 when a client asked for an explanation of a systematic difference between the mass of metal (Pb) entering his plant and the mass of metal leaving it as residues and marketable concentrates. After a long and detailed study the most serious error detected was a bias of 0.9% in the dynamic weightometer on the feed belt to the plant. The apparatus was calibrated every morning by a heavy roller chain of known length and mass simulating a given flow-rate in a way that was itself wrong, and so it artificially biased the recorded input.

Since then there have been many calls to resolve similar problems. It may be noted that the observation made in 1955 on dynamic weighing is completely general, an opinion shared with other specialists on the subject of weighing. In a book written in 1975 [18], H. Colijn wrote:

> The actual plant performance of belt scales, unfortunately, does not always measure up to the claims of the manufacturer or to the expectations of the operator. Instead of the $\frac{1}{2}$ percent accuracy, some plant operating personnel have claimed that 10 percent is a more realistic figure, and on a large number of installations they may be correct. As a result, many belt scales are either used for only rough approximations or allowed to sit completely idle. Many operators have become very discouraged and apprehensive about the use of belt scales at all.

Subsequent discussions with H. Colijn have confirmed that the opinions he expressed in 1975 on mechanical integrating weighing machines have been, if anything, reinforced further.

As for nuclear weightometers, he noted that their manufacturers claim a 'precision' of ±1% but indicated that this figure is undoubtedly optimistic. The author has been able to verify this for himself on many occasions.

Note: Like some of his colleagues, and likewise in some of the standards, H. Colijn appears to confound the concepts of accuracy and precision, subjects about which there seems to be complete confusion. The evidence suggests that this results from failing to use clear, unambiguous, scientific definitions such as those given in Section 4.5 of this book and in Sections 1.7.2 of references [6] and [7]. In fact English-language authors use the word 'accuracy' in two different senses. The confusion was underlined by the statistician J. Mandel (*The Analysis of Experimental Data*):

> Regarding the concept of accuracy there exist two schools of thought. Many authors define accuracy as the more or less complete absence of bias. The second school of thought defines accuracy in terms of the total error [. . .]. For the sake of consistency we will adopt the first view only.

This was a conclusion the author came to independently of Mandel and which he adopted too, even though the second definition is that used by the British Standards.

> *Example*. British Standard BS 1017: 'The sampling of coal and coke': 'Accuracy is the measure of the ability of a method to provide accurate results, i.e. results which are precise and free from bias.'

Some confusion also exists between the French word 'precision' and the English word 'precision'. This is why a scientific definition has not been given to either the French word or the English one. When the word 'precision' is used it is in the sense of the broad, vague, non-scientific usage of everyday language, as, for example, when the expression 'precision instrument' is used to describe some undefined quality that includes accuracy and reproducibility. Indeed, both languages need some word of a suitably non-specific nature.

Regarding the ways in which the volumes of liquids are measured by mechanical meters or flowmeters, it will be seen in a following section that these give cause for concern. Flowmeters, electromagnetic or otherwise, create unease. For example, in one plant two flowmeters of the same make and model were installed in series in order to detect the causes of a bias. Despite being calibrated daily by a graduated 5 m^3 tank so that the raw readings could be corrected, these two identical instruments continued to give readings that differed systematically by several percentage points. There was no possibility of recommending them.

The mass of solids in a flowing current of pulp can be estimated, if the solids are finely ground, by means of integrating electromagnetic flowmeters that estimate (though poorly) the volume, and of gamma-densimeters that convert (hardly better) the volume into the mass of pulp. Neither of them is to be recommended.

The gravest fault of all the indirect measuring appliances is that they invariably give numbers that are stored away somewhere, or more recently stored automatically in a computer's memory bank, without even the most cursory critical examination first having been made of their validity or reliability. The reader will have noted that, in every example given above, the estimates of masses, volumes and densities are made from indirect measurements. Everything else but masses and volumes are measured, and the mathematical relationships between the magnitudes of the measurements and the unknown masses or volumes to be estimated very often need to be treated with caution.

One of the merits of proportional sampling is that the operator is required to estimate moving masses by means of a balance or a weighing machine and volumes by graduated receptacles.

13.2 PRINCIPLE AND ACCURACY OF PROPORTIONAL SAMPLING

The sampling rates τ and τ' were defined in Section 2.1 as:

$$\text{The mass sampling rate } \tau = \frac{\text{Mass } M_E \text{ of the sample } E}{\text{Mass } M_L \text{ of the lot } L}$$

and

The time sampling rate τ'

$$= \frac{\text{Time during which the } Q \text{ increments } I_q \text{ making up } E \text{ were taken}}{\text{Total time during which the lot } L \text{ was flowing}}$$

The time sampling rate τ' can be measured or calculated with excellent 'precision' from the design of the sampling device.

The mass of the sample M_E is a random variable. From the definition of τ given above, and from theoretical considerations (see Section 29.3.3 of reference [7]) it follows that

$$m(\tau) = m(M_E)/M_L; \quad m(\tau) = \tau' \quad \text{so that} \quad m(M_E)/M_L = \tau'$$

whence:

$$m(M_E)/\tau' = M_L \text{ and } M_E/\tau' = M'_L = \text{the estimator of } M_L$$

The estimator M'_L is itself a random variable with mean:

$$m(M'_L) = m(M_E)/\tau'$$

The mass M_E, which is small, can be weighed on a 'precise', reliable static weighing machine that can be calibrated easily. The time sampling rate τ' can easily be determined from the design of the sampling device, as will be seen below.

When sampling is correct and therefore accurate, but only then, M'_L is an unbiased estimator of the mass M_L of the lot L. The mass of the sample is structurally proportional to that of the lot sampled.

A structurally unbiased estimator! Not one of the integrating weighing machines of any kind examined worldwide has been found to be capable of meeting this requirement.

The technique of proportional sampling is founded on this unique property

In practice lot L is sampled correctly with a uniform selection probability P, which, by definition, is identical to τ', by taking a large number Q of small increments I_q that together make up the sample E.

To be reliable, correct and therefore accurate 'proportional' sampling must have, in addition, the property of excellent reproducibility, which in turn demands a very low coefficient of variation in the masses of the increments. This point is covered in Section 13.3 below.

The coefficient of proportionality $\tau' = P$ must be known with a high degree of precision and remain stable in time, a point covered in Section 13.4 below. In practice, time sampling rates are of the order of 1% to 0.1% with continuous samplers, which are known to be the most precise machines. Lower sampling rates (with no lower limit but with a corresponding reduction in their reproducibility and reliability) can be had with discontinuous samplers.

It is enough, therefore, to weigh the sample by means of a static weighing machine, and then divide the mass M_E by the time sampling rate τ' in order to obtain an unbiased estimate M'_L of the mass M_L of lot L.

The observations made about masses apply, *mutatis mutandis*, to volumes as well; liquids can be measured more precisely by weighing them.

13.3 THE THEORETICAL REPRODUCIBILTY OF PROPORTIONAL SAMPLING

Let M_i be the mass of the constituent element F_i in the stream and extend the sum $\sum$ to encompass all the N_F fragments, molecules or ions F_i making up lot L. If sampling is correct and only then, from equations 17.20 and 17.21 of Section 17.3.6 of reference [7]:

$$P_m = P = \text{constant}$$

M_E is then a normally distributed random variable:

- With mean: $m(M_E) = P\sum M_i = PM_L$

- And variance: $s^2(M_E) = P(1 - P)\sum M_i^2$

The relative standard deviation of M_E can be written:

$$\frac{s(M_E)}{m(M_E)} = \frac{(1 - P)^{1/2}}{P^{1/2}} \cdot \frac{(\sum M_i^2)^{1/2}}{\sum M_i}$$

In the left-hand term of the product, the probability P is of the order of 1% to 0.1%, or 0.01 to 0.001. Thus, the fraction $(1 - P)/P$ takes values between 100 and 1000 and its square root lies between 10 and 32.

To get some idea of the order of magnitude of the second fraction of the product, it will be assumed that all the N_F elements F_i have the same mass, equal to M_L/N_F. The numerator can be written:

$$[\sum M_i^2]^{1/2} = [N_F(M_L/N_F)^2]^{1/2} = M_L/N_F^{1/2}$$

The denominator $\sum M_i$ is none other than M_L, so the quotient simply reduces to $1/N_F^{1/2}$.

In the case of crushed solids, or to an infinitely greater degree of liquids, N_F is always a very large number. Consider a somewhat unfavourable example: that of a lot of 100 tonnes of 1 cm diameter spherical granules of density 2.5 g/cm^3. Assuming a shape parameter of 0.5 (see section 8.3.4), the volume of each fragment is 0.5 cm^3 and it weighs 1.25 g. N_F is equal to 80 million and the reciprocal of its square root is 0.00011. If the sampler has been set to a time sampling rate of 1% (0.01), it is easy to calculate that:

$$s(M_E)/m(M_E) = 10 \times 0.000\,11 = 0.0011$$

This value represents an extremely small relative dispersion.

In the case of an aqueous solution, knowing that there are 6.10^{23} molecules in 18 g of water, or 33.10^{21} molecules per gram, then for a lot of 100 tonnes $N_F = 3.3.10^{30}$.

The quotient $1/N_F^{1/2}$ becomes $0.55.10^{-15}$ and, for $P = 0.01$, the coefficient of variation s/m is 5.10^{-15}, which corresponds to a theoretical reproducibility of astronomical proportions. However, there is every reason to believe that operating errors will be very much greater.

These calculations confirm the belief that proportional sampling is a tool for measuring moving streams of solids and liquids that not only is unbiased but also has extremely good reproducibility. These qualities are far beyond the belt scales and dynamic volume measuring meters currently in use.

To summarise, the unbiased estimator M'_L of the mass M_L of lot L can be written

$$M'_L = \text{unbiased estimator of } M_L = M_E/\tau'$$

The mass M_E can be determined accurately and reproducibly. All that now remains is how to estimate τ'.

13.4   ESTIMATION OF THE TIME SAMPLING RATE τ': CONTINUOUS SAMPLING SYSTEMS

Only transverse linear sample cutters and rotating sample cutters are able to ensure that proportional sampling is done absolutely correctly. The type of continuous straight-line sample cutter (i.e. not coming to rest at the end of each cut) considered here is the simplest: the one operating with an endless chain as shown in Figure 13.1.

The following are defined:

C the length of the endless chain (in m)
W the width of the sample cutter (in m)
V the speed of the lug, and hence of the cutter.

The continuous running of the geared motor guarantees that the speed of the cutter during its travel across the stream is for all practical purposes uniform. This is especially so if the motor is overdesigned to prevent the cutter slowing down as it enters the stream and comes under load (as it may do in the case of high flow rates).

This is an important consideration that is often overlooked by the makers of transverse samplers. The author was called on to design a plant to sample a stream of iron ore delivering 16 000 t/h, with peaks of 20 000 t/h (or 5.5 t/s),

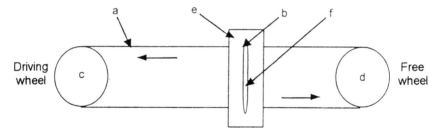

a endless chain
b lug fixed to chain and sliding with minimum play along slit (f)
c cog wheel driven by reduction motor
d idler cog keeping the chain in tension
e metal frame constrained to move smoothly and parallel to itself
   from left to right and from right to left ; the cutter is fixed to this
   as firmly as possible.
f slit with a width slightly greater than the diameter of lug (b)

Figure 13.1   Schematic representation of a continuous transverse chain sampler

carried on a moving belt and discharging into an ore-carrier. The belt was 3.3 m wide and travelled at 4 m/s. The calculation of the kinetic energy of the moving stream of ore and the force with which it struck the sample cutter is left to the reader. This is not the place to describe the installation which, as far as is known, holds the world record in this respect, but in agreement with the electrical engineer after having carefully calculated the minimum power needed by the main motor ... it was generously doubled so as to avoid any unpleasant surprises and to be able to guarantee the uniform speed of the cutter as it traversed the stream, and hence ensure the correctness, which was vital, of commercial sampling.

The speed V of the sampler enters only to the extent that it must respect the following two rules for correct sampling (Section 6.2.7):

- On one hand, it must be uniform and
- On the other, it must not exceed 0.4 m/s (for proportional sampling the normally recommended maximum cutter speed of 0.6 m/s is reduced by one third so as to introduce a safety factor).

Therefore, whatever the cutter speed V, the proportion of the time during which each element of the section of the current is traversed during one pass of the cutter is equal to W/C. During one cycle (= time taken by the lug to make one complete tour) the cutter cuts the current twice: once from right to left and then once from left to right, and so the time sampling rate is

$$\tau' = 2W/C$$

To use this expression in practice requires a precise knowledge of the length C of the chain and the width W of the cutter. The length C is easy to measure when new. The links usually have a standard length and it is easy to count them. In one new installation, for example, C was 90 inches or 2.286 m long. In practice, maintainance technicians must check the chain periodically to ensure that it is not worn out, and if need be to change it. If worn, it is recommended that the chain is changed to avoid uneven movements or sudden jerks; this is recommended despite the fact that provided the drive wheel rotates at uniform speed it is the number of links that counts, not their length. The sampling rate τ' remains unchanged in spite of wear.

Regarding the effective width W of the cutter slot, the problem is rather more complicated than appears at first sight. Up to now it has been assumed that the width of the cutter is delineated by two infinitely thin parallel plates. In reality they are never as thin as razor blades, but at best comprise two sheet metal plates of a constant, non-zero thickness e. Figure 13.2 shows the arrangement. Generally, the thickness e of the leading edge of the cutter is small in comparison with the distance W', assumed constant, between the internal faces of the two plates.

The following simplifying hypotheses are made:

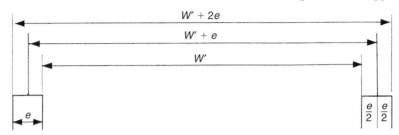

Figure 13.2 Nominal aperture W' and effective aperture W of a sample cutter with two parallel plates. Vertical section

1. The stream sampled is pure water,
2. Stream flow is perfectly laminar. Experience has shown, repeatedly, that this condition is practically a *sine qua non* for reliable proportional sampling,
3. The water leaving the cutter does so without overflowing, splashing or being impeded.

In these conditions it is possible to assume further that when the stream of water strikes one of the two cutter walls it is split symmetrically, as if the top of the wall were a prism with the cross-section of an isosceles triangle. Thus:

$$W = W' + e/2 + e/2 = W' + e$$

When the sampled material is not water and when the stream flow is turbulent, as long as condition 3 above still holds good it remains valid to write:

$$W' + 2e >= W >= W' \text{ so that: } W = W' + e \pm e$$

It is clear, therefore, that the relative uncertainty $e/(W' + e)$ can be reduced by increasing W', which guarantees extra correctness—but at the cost of increasing the mass of the sample—and by reducing e without weakening the cutter. A compromise has to be made in every case.

Figure 13.2 represents the tops of the cutter plates as having sharp angles, but these are soon rounded off by wear and tear, a fact that tends to increase the validity of taking the value of W as the effective opening.

In the case of a continuous rotary sampler in which the cutter edges form an angle $\alpha°$ at the centre, the time sampling rate is obviously equal to:

$$\tau' = \alpha/360$$

Estimating the angle α calls only for some simple trigonometry.

What is of prime importance is that the arms of the cutter pass through the axis of rotation.

13.5 INDUSTRIAL-SCALE AND PILOT-PLANT REGIMES

It is not expected that readers will be entirely convinced by what has been written so far, solely on the basis of theoretical development and personal conviction. Accordingly they should be told that this original technique has, since being recommended in 1980, been in operation on an industrial scale in several plants treating platinum ores in South Africa at Rustenburg Platinum Mines, where it was used to calibrate nuclear scales installed at the time the plant was constructed and which were giving conflicting results. In 1992 the plants were running smoothly and operating satisfactorily.

> *It seems unnecessary to state that the world's leading producers of platinum and associated metals would be extremely demanding of anything that affected the control of their plants and the recovery of metal.*

One variation of the technique described above consists, in a pilot plant, not of calculating the time sampling rate τ' but of ensuring, by construction, that it is strictly the same for all the material flows circulating in the plant. The masses of the samples collected and those of the lots they represent are in the same proportion, and this allows the plant's mass balance to be calculated very simply, with the added bonus of gathering samples for analysis.

This technique has been used by the BRGM (Bureau de Recherches Géologiques et Miniéres at Orleans, France) for more than twenty years and with complete success. (The author wishes to thank the directors of Rustenburg Platinum Mines and of the BRGM for giving permission to quote them in his publications. Particular thanks are given to Dr Hugh Bartlett and to Alain Broussaud.)

13.6 ESTIMATION OF THE TIME SAMPLING RATE τ': DISCONTINUOUS SAMPLING DEVICES

Up to 1992 [6, 7] the applications of proportional sampling have been limited to continuous samplers, thereby achieving higher sampling rates. Since then studies have been and continue to be made of proportional sampling using either continuous or discontinuous samplers that have much lower sampling rates. No more will be said of these appliances, which are of various types, as they are presently being patented.

13.7 AN EXPERIMENTAL STUDY OF PROPORTIONAL SAMPLING

The reader may not be convinced by the arguments advanced so far, because although Rustenburg Platinum Mines authorised a description of the original technique they did not allow the publication of the results obtained from their plants. There is nothing novel about this attitude in industry.

Sampling philanthropists being an extremely rare species who are in need of protection from the WWF, advantage was taken of an offer made in 1984 by a well-intentioned company who allowed some tests to be made on the 'calibration' of proportional sampling in their pilot plants, which were running smoothly.

The idea was, of course, to detect, using whatever weighing machines were available in the plants, any possible faults or failings in proportional sampling. The experiment turned out to be an amusingly modern version of the biter bit, an outcome foreseen by no one. The installation is described in Figure 13.3.

The sampler used was of the endless chain type illustrated in Figure 13.1. From its dimensions it was calculated, before the test began, that the time sampling rate τ' was 1.937%. A high degree of 'precision' was expected and so expressing the rate to four significant figures did not seem out of place.

In all the sampling tests the cutter (g), fed through the overflow tank (f), sampled the stream delivered from the tank (a) by pump (e). Successive samples were collected in the previously weighed containers (h). To interpret the masses shown by the digital display, the convention was adopted whereby the mass of the tank was equal to M at the exact moment when the display changed from $M + 20$ to M kg. When the tank was full the meter showed abut 10 200 kg. The pump was started and the experiment proper began when the display showed exactly 10 000 kg.

1. *First test of proportional sampling*: this was run as a familiarisation exercise. Being winter, the temperature of the air and water were about 6°C. Some gusts of wind were noted (the tank was in the open air). The feed rate varied from 7.5 t/h at the beginning of the test to 4.3 t/h at the end. The tank was gradually run down from 10 000 to 1000 kg, while the collection vessels were changed at precisely the end of each 1000 kg of water delivered. The object of the trial was to compare the observed mass sampling rate with the calculated time sampling rate. The mass rate peaked at 2.02% (the first delivery of 1000 kg) and at 1.96% (the ninth and last delivery). Between 9000 and 2000 kg it oscillated between 1.94 and 1.95%. The theoretical rate, by calculation, was 1.937%. The two sets of results would be consistent if two high sampling rates were omitted. Two explanations could be advanced.

The first could be some abnormal behaviour in the sampling system. By inspection, it seemed impossible that anything other than a negligible bias could be present. The second could be a fault in the weighing system.

Note: the experimental control of the accuracy of proportional sampling only makes sense to the extent that the system of control is faultless. In other words, it makes no sense to check the length of a standard metre rule with a much-used linen tape measure. The author's confidence in proportional sampling, which relies largely on its simplicity and transparency, led him to question the weighing system in this case.

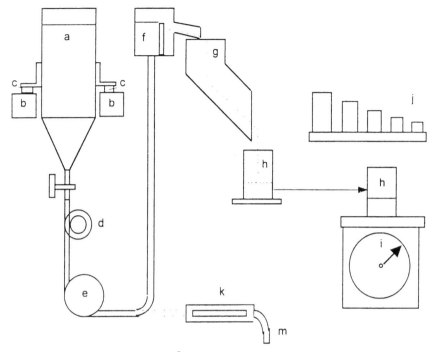

a cylindro-conical tank of 10 m^3 capacity
b ring girder supporting the tank
c set of three strain gauges reading digitally in 20 kg increments
d flexible looped tube allowing the tank to empty independently of the pump
e centrifugal pump
f feeder box regulating the flow rate
g transverse increment cutter
h several containers each holding 60 litres
i 60 kg balance weighing to the nearest 20 g
j set of standard weights
k water meter
m drainage channel

Figure 13.3 Control system for the proportional sampling experiment

2. *First recalibration of the measuring system.* Calibration employed the following instruments:

- A set of three strain gauges with a digital display reading in increments of 20 kg
- A water meter graduated in increments of 0.1 litre
- A 'Testut' balance, range 60 kg, readable to the nearest 20 g
- A set of standard weights of 10, 5, 2, 1 and 0.5 kg.

(a) *Balance*: it was observed that at no position on the dial, including the zone below 5 kg, was the difference wider than the thickness of the pointer. It was considered that the balance was safe to use.

(b) *Water meter*: ten tests were made, each one by weighing 50.0 litres of water to the nearest 20 g; the volumes were measured by differences between successive readings. At 6 °C the mass of 50.0 litres of water does not differ significantly from 50.0 kg. The results showed a mean mass of 49.6 kg. Emptying the tank by gravity took two days, and the same tests made on the following day while the tank was being emptied gave a mean mass of 50.05 kg. No explanations could be found to account for these differences.

(c) *Weighing strain gauges*: it was decided to measure the volumes of consecutive 500 kg batches (as measured by gauge) by means of the water meter. A major anomaly was detected in the first delivery. The meter showed a volume of 538.5 litres, a bias of 8%. The volumes of the other 500 kg deliveries varied between 492 and 506 litres (the last). The meter, although itself raising some doubts, seemed to confirm the two anomalies detected by proportional sampling at the beginning and end of the first trial.

(d) *Conclusions from the first trial of proportional sampling*: Only limited confidence could be placed on the methods of determining masses and volumes.

3. *Trials numbers 2 to 5 on proportional sampling*: these involved several modifications made to the feed box to ensure, as far as possible, laminar flow or at least a flow disturbed to the least extent possible by turbulence. One conclusion was that the flow rate had been too high, and this was reduced to a mean rate of 5.8 t/h.

4. *Trial number 6 on proportional sampling*: The flow was, by inspection, as laminar as possible, thanks to modifications in the feed box and the reduction in the flow rate. The mean mass sampling rate (over 9000 kg as measured by the gauges) became 1.934% compared with the time sampling rate of 1.937%. However, it was too early to claim success, because although the mean was acceptable a serious dispersion was found between the 1000 kg batches confirming, as had been suspected already, that the weighing system contained some anomalies. After reflection and observation several possible causes of malfunction were listed.

The first was that the tank was not completely free from the circular girder supporting it (it should have been resting on the strain gauges only): several points of contact were found and subsequently removed. The second was that the water meter gave unsatisfactory readings: it was stripped down completely, its parts cleaned of accumulated dirt, oiled, and its cog wheels checked to ensure they were working properly. Finally, it was recalibrated.

5. *Calibrating the water meter by the spring balance*: Despite the above precautions, some puzzling results ensued. It will be recalled that it took two days to empty the tank by gravity (10 000 to 1000 kg). Repeated weighings by the balance of batches of 50.0 litres measured by the water meter, made during the morning of the first day between 9 and 11 a.m., indicated a perfect accuracy

of 50.0 kg for 50.0 litres. But that didn't last. From 11 a.m. to 4 p.m. batches of 50.0 litres (by water meter) weighed 50.03, 50.04 and 50.08 kg. On the second day weights of 50.16 and 50.26 kg were recorded. The temperature, which did not vary, could not have been the cause.

It was observed that a water meter made by a reputable firm, well maintained, not showing any apparent faults and up to that point being perfectly satisfactory, began to show a drift after some hours of operation, even when only pure water was flowing through it. This reinforced the original view that the measurement of volumes of liquids, even ones as neutral as pure water, is not done as accurately and as reproducibly as the makers believe.

6. *Strain gauge calibration by using the water meter*: to check weighings made by the strain gauges against a water meter that is known to be lacking in reliability seems to be illogical, but it was believed that errors in weighings were an order of magnitude larger than those made by the water meter. It was planned to measure the volumes of consecutive 500 kg batches (as measured by strain gauge) by means of the water meter, but they were reduced to 100 kg and even down to 20 kg between 4500 and 1000 kg after the results became inconsistent. It was observed that the two sets of measurements had been in perfect agreement between 10 000 and 4500 kg. Then, things went wrong and the observations became erratic, so it was decided at that point to measure every 100 kg and then every 20 kg.

Between 4000 and 3500 kg (by weighing), the water meter should have registered 500 litres but showed 580.7 litres, a relative bias of 16%; between 3000 and 2500 kg: 437.5 litres instead of 500; between 3800 and 3700 kg: 161.8 litres instead of 100; between 2800 and 2700 kg: 64.3 litres instead of 100; between 3740 and 3720 kg: 66.3 litres instead of 20; between 2660 and 2640 kg: only 9 litres. The flow rate of the water leaving the water meter had been timed, albeit approximately, and had seemed to be perfectly uniform. The water meter could not have been the cause of such deviations.

Therefore, all doubts were centred on the weighing system. It will be recalled that the appliance, a well-known make, had worked 'to everyone's satisfaction'! There could be no question that proportional sampling which has a reproducibility of the order of 0.1% could be checked by such an appliance. This agreed with the observations, unfortunately qualitative only, of other clients. The trials led to the assumption, but not to the proof, that proportional sampling does better than static weighing. In the case of dynamic weighing it is known that the South African mines have irrefutable numerical data proving the reliability of proportional sampling which also gives, for good measure, samples for analysis.

13.8 PRACTICAL REPRODUCIBILITY OF PROPORTIONAL SAMPLING

This was estimated as follows. Fifty flasks of 500 ml capacity were weighed to the nearest 0.01 g and numbered from 1 to 50. The flasks were placed in order and used to collect successive increments taken at a uniform interval of 2.68 seconds. After some familiarisation trials, the cut-off point between one increment and the next could be fixed to the nearest drop. The 50 increments were then weighed in the same order to the nearest 0.01 g. Two flasks were found to contain fragments of rust and were discarded. The population of the remaining 48 increments had the following moments:

- Mean mass 96.2058 g
- Absolute standard deviation 0.7352 g
- Relative standard deviation 0.76%

A variographic analysis of the 48 masses gave a flat variogram. The feed rate was quite constant as the whole operation took only 134 seconds.

13.9   HYPOTHETICAL APPLICATION TO A LOT OF 10 000 KG OF WATER

Suppose that proportional sampling was used, under conditions based on previous experience, to estimate the mass of a lot of 10 tonnes of water flowing in 90 minutes. The sample so recovered would be made up of 2014 increments. The distribution of the mass M_E of the sample E would then have the following moments:

- Mean $m(M_E)$ 193.745 kg
- Absolute standard deviation $s(M_E)$ 32.994 g
- Relative standard deviation s/m $1.7.10^{-4}$

which is far better than anything the designers of current sampling systems claim they can offer and infinitely better than their actual performance, which was estimated by Colijn to be some 10% in error. Even though this example is hypothetical it gives a clear indication of the difference between proportional sampling and what is currently available. If a time sampling rate of 1.937% is given to this proportional sample, it can be calculated that for a mass M_L of a lot L the unbiased estimator M'_L has moments:

- Mean $m(M'_L)$ 10 002.3 kg
- Absolute standard deviation $s(M'_L)$ 1.7 kg
- 95% confidence limits ± 3.4 kg

Thus, the estimated value of a mass of 10 000 kg is:

$$M_L = 10\,002.3 \pm 3.4 \text{ kg}$$

The maximum deviation from reality (10 005.7 kg instead of 10 000 kg) is 0.057%, which is excellent.

13.10   IN CONCLUSION

It can be seen that proportional sampling is an extremely promising technique, one that may revolutionise measuring the masses and volumes of flowing streams of solids, liquids and multi-phase materials. The modest experiments described above have suggested, although they have not as yet definitely proved, that the technique of proportional sampling is accurate and reliable. More experiments need to be done with the object of providing unassailable proof. Trials have shown that static strain gauges and a water meter could, in contrast to proportional sampling, give aberrant results. The author's experience as a consulting engineer has shown him that there are many industrial users who complain of a lack of reliability in the instruments they are using to measure the masses and volumes of flowing streams by indirect methods. H. Colijn, a specialist in industrial weighing methods, agrees with these views [18].

Proportional sampling has two attractive characteristics. The first is its simplicity and transparency: if elementary precautions are taken to prevent foreign materials obstructing the sampler; if regular checks are made to ensure that those two critical values, the width of the cutter slot and the length of the sampler chain, are maintained properly then simple visual observation is all that is needed to check that the sampler is working properly. It can be added that all other methods of weighing or of measuring volumes, of the 'black box' type, can produce any number of deviations of which the user is totally unaware.

It is common practice—and a very convenient one—to assume dogmatically that a linear relationship exists between the value of x that needs to be estimated and a value y that is actually measured, which is very often some electrical characteristic that has no simple relationship with x; a glance at Colijn's book [18] will convince anyone on this point. If linearity can be proved all that needs to be done is to calibrate two points and draw a straight line. But such linearity— which is also found in many types of analysers—is more often a matter of faith (or of hypothesis or wishful thinking) rather than of scientific proof.

The second is that proportional sampling, done absolutely correctly, gives ideal samples for laboratory analysis. Thus, proportional sampling kills two birds with one stone: the same appliance provides industry with all the quantitative and qualitative information that characterises a stream of material. Economy and simplicity combine.

It is obvious that the system should be the responsibility of the quality control service and that person should be advised by a specialist whose education has included a proper course on the science of sampling. In conclusion, it is hoped that everyone in charge of quality control services takes heed of the merits of sampling theory in general and of proportional sampling in particular.

The only fault of proportional sampling is that it is simple; too simple in an age when 'sophistication' is all.

Appendix 1: A Brief Résumé of Classical Statistics (Non-weighted)

This appendix is not intended to be a complete course in statistics, but simply serves to remind readers of some concepts and definitions that they may find useful while reading this book.

THE CONCEPTS OF OBJECTS, CONSTITUENT ELEMENTS, SETS, POPULATIONS AND SERIES

Statistics deal with assemblages of 'objects' (in the broadest sense) that can be either material, such as the fragments comprising a lot of broken material, or immaterial such as the analytical results obtained from a particular sample. From a mathematical point of view these assemblages of objects are called 'sets'. Here, we are dealing with two kinds of sets: 'populations' and 'series'. In a 'population' the objects are not presumed to be ordered. Even if some order did exist it would not be taken into account. In a 'series' the objects are ordered, usually on a geometrical or chronological basis. In the latter case one often speaks of 'time series'.

The 'constituent elements' making up a population of material objects can have the same physical mass, e.g. closely calibrated fragments; they may then be given the same 'statistical weight'. When they have different masses, e.g. non-calibrated fragments, they must be given 'statistical weights proportional to their masses'.

Classical statistics, taught ubiquitously, applies only:

- To populations of non-ordered objects and
- To objects having the same mass and therefore the same statistical weight.

It is a mathematical mistake, a grave one in practice, to apply its laws:

- To sets (populations or series) of objects having different masses
- To series of ordered objects irrespective of their masses

Appendix 2 gives a schematic illustration of the three branches of statistics:

- Statistics of populations of objects with the same weight (or 'classical statistics')
- Statistics of populations of objects with different weights
- Statistics of series of objects, e.g. chronostatistics.

CONCEPT OF A SAMPLE

Whether the lot of interest is material or immaterial, its evaluation can never, mostly for reasons of economics, be based on the analysis (in the broad sense) of the whole set of its constituent elements. It has to be performed on a fraction of the whole, a sub-set of these elements that has to be selected adequately. Such a sub-set is called a 'sample'. When the selection is done inadequately this sub-set is called a 'specimen' and must be regarded as unreliable. The conditions of an 'adequate' selection are set out precisely in Chapter 4.

CONCEPT OF RANDOM VARIABLES

Much can be written about chance and the adjective 'random' but this would launch a metaphysical discussion that would be out of place here. Nevertheless, it can be noted that in populations of material units the properties of each unit are perfectly defined and owe nothing to chance. On the other hand, in populations of immaterial units, such as the results of repeated analyses, a large number of small causes act together to make each result different from the others.

Let x be some quantitative property of a material object or the value of an immaterial object. If x has the properties of a random or a quasi-random variable it can be characterised, on the one hand, by its 'law of probability or distribution' and, on the other, by what are called its 'moments'.

LAW OF PROBABILITY OR DISTRIBUTION

There are many of these but most have no connection with the present work. Only two kinds of random variable are of interest here:

- The 'analytical grades' of lots of material and of the samples representing them (well or badly—the whole object of this study)
- The 'errors of sampling and analysis'.

The distribution of the analytical grades of the units that constitute a lot of material can follow any kind of law as they are the results of deterministic phenomena. The ones of interest here are essentially the distributions of sample grades and of the errors of sampling or analysis. These distributions almost always obey a law known as Gauss's or Laplace–Gauss's law (although it was in fact discovered by de Moivre in 1753). It is called more simply the 'normal' distribution law, which greatly simplifies the problem of its paternal attribution.

The normal distribution, known universally, is characterised by the 'Gaussian curve' shown in Figure A1.1. The standard deviation *s* is defined later.

The exception justifying the adjective 'almost' arises from very low grade materials in which the analyte is present in trace amounts. This point (What is low-grade material? What are traces?) is gone into in greater depth in the main texts (see References). Their distribution is asymmetric, whereas the normal one is symmetric; they tend to be 'log-normal', in that the logarithms of *x* tend to follow a normal law.

THE CENTRAL LIMIT THEOREM OR THE THEOREM OF LAPLACE–LIAPOUNOFF

This is of prime importance in sampling theory as it states that whatever the distribution law of the parent population, even an extremely asymmetric one, the distribution law of the means of increasingly large samples drawn from it approaches ever more closely that of a normal distribution (a sample of infinite size merges into the parent population itself). This condition is always satisfied in the case of material units and the same remarks hold for errors as well. For all practical purposes, with the exception of small samples in which the analytes are present as traces, the distributions can be regarded as normal.

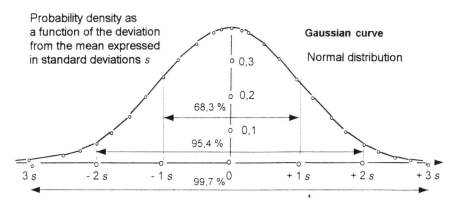

Figure A1.1   Illustration of the Gaussian distribution curve with its familiar bell shape

THE 'MOMENTS' OF A DISTRIBUTION

These are the characteristics of a population that are specified or calculated from the given set of values of the random variable. The moment of order 1 is the arithmetic mean, or mathematical expectation, or simply the mean. It characterises the central tendency of the distribution. If the variable x is discontinuous and can take N discrete values $x_1, \ldots x_n, \ldots x_N$, its mean $m(x)$ can be written:

$$m(x) = \frac{1}{N} \sum_{n=1}^{n=N} x_n$$

For moments higher than one, centred moments, which characterise the dispersion of the values of x about their mean $m(x)$, have to be distinguished from non-centred moments. Moments of orders higher than 2 are of no interest here.

The centred moment of order 2 is called the variance. It characterises the dispersion of the values of x and is written $s^2(x)$ or $\sigma^2(x)$. $s^2(x)$ will be used for preference.

$$s^2(x) = \frac{1}{N} \sum_{n=1}^{n=N} [x_n - m(x)]^2$$

The square root of the variance is the standard deviation $s(x)$, or s as illustrated in Figure A1.1.

In the normal distribution the value of the variance is independent of the value of the mean. That is why neither the mean nor the variance is by itself able to characterise a normal distribution completely. Some global characteristic is needed that includes both the mean and the variance.

The non-centred moment of order 2 is called the mean square. It is denoted by $r^2(x)$. It is the mean of the squared values of x_n:

$$r^2(x) = \frac{1}{N} \sum_{n=1}^{n=N} x_n^2$$

It can be shown that

$$r^2(x) = m^2(x) + s^2(x)$$

The mean square $r^2(x)$ encompasses both the mean and the variance, so serving the purpose of characterising the distribution 'globally' (see Section 4.5). The definition of 'representativeness' rests on this statistic.

PRINCIPAL PROPERTIES OF THE NORMAL LAW

The curve shown in Figure A1.1 shows the 'probability density' as a function of deviations from the mean that are expressed in terms of standard deviations, as the reader well knows. It is symmetrical, implying that there are equal chances of finding the same numbers of positive and negative deviations.

In practical terms, what is to be gained by assuming that a particular distribution is normal? One application is to be found in the confidence limits associated with estimates of grades and errors, as defined next.

NORMAL LAW: CONFIDENCE LIMITS

Assume that, as the result of a non-biased estimation process, it has been decided that the resulting distribution is normal, or nearly so. The mean m, the variance s^2 and the standard deviation s have been calculated, from which:

- The interval $m \pm 1s$ contains 68.3% of the individual values
- The interval $m \pm 2s$ contains 95.4% of the individual values
- The interval $m \pm 3s$ contains 99.74% of the individual values

The confidence interval $m \pm 2s$ is often used to signify a serious presumption that the actual unknown value lies between these limits (less than 5% or only one chance in 20 of being wrong), while the interval $m \pm 3s$ implies a virtual certainty (only 0.26% or one chance in 385 of being wrong). These confidence intervals are based on the assumption of a perfectly normal distribution. In practice, the percentages listed above are regarded simply as orders of magnitude.

A point of language: one cannot say that the actual value, always predetermined but nevertheless still unknown, has a certain probability of lying within a particular confidence interval, because the actual value is not a random variable. It can only be said that the confidence interval has a certain probability of bracketing it.

DEFINITIONS OF A SAMPLING PROCESS

Section 4.5 defined a sampling process as:

- *Exact*: a property of the error itself, which must be identically zero
- *Strictly accurate*: a property of the mean which must be identically zero.

These two definitions correspond to nothing in the real, material world. Instead, a sampling process is defined as:

- *Accurate in practice*: characterised by a mean that is not strictly zero but is instead equal to or less than, in absolute terms, some acceptable maximum that usually has a negligible value
- *Reproducible*: when its variance is less than or equal to some acceptable maximum

- *Representative*: when its mean square is less than or equal to some acceptable maximum.

In sampling and analytical processes a distinction can be recognised between the repeatability (of results obtained under identical conditions from the same series of analyses in the same laboratory) and the reproducibility observed when at least one of the above conditions is not satisfied. In terms of variances, repeatability is the absolute minimum reproducibility.

GRAPHICAL ILLUSTRATION OF THE ADDITIVITY OF BIASES AND VARIANCES

First example

It will be assumed that the primary and secondary sampling stages and the analysis are accurate; that is, unbiased, and each distribution is normal. This case is illustrated in Figure A1.2. Because the three operations are unbiased it follows that (in the notation of Section 1.8.1):

$$m(a_R) = m(a_{E1}) = m(a_{E2}) = a_L$$

The variances of the three random operations are additive, and because of this the first Gaussian curve from the primary sampling stage is less spread than the second, which covers both primary and secondary stages, and this in turn is less spread than the third that covers primary and secondary sampling plus analysis. Each curve is more spread than the preceding one because of the cumulative

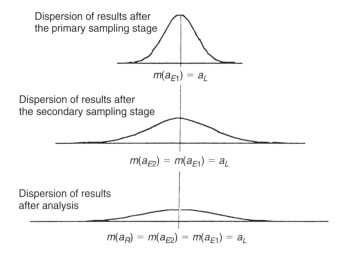

Figure A1.2 Spread of the analytical results when the primary and secondary sample preparation stages and the analysis are all unbiased

effect of the preceding operations. All three curves tend asymptotically to zero on both sides. By definition, their integral (the area between the curve and the abscissa) is equal to unity.

The analytical result a_R is an unbiased estimator of the actual value a_L. The distribution of the analyses is normal and its mean $m(a_R)$ is identically equal to the true unknown value a_L of lot L which one is trying to determine.

Second example

The first and second sampling stages and the final analysis are all biased (Figure A1.3). The final result a_R is a biased estimator of the true grade a_L. The bias is equal to b_G, but usually it is unknown. This is why it is essential to eliminate, or at least to minimise, any bias.

The lowest curve in Figure A1.3 shows that the deviation between the true value a_L—indicated by the vertical broken line along the abscissa a_L running from top to bottom of the figure—and the mean $m(a_R)$ of the final result corresponds to a shift of the order of $-3s$, i.e. three standard deviations. That is, the confidence interval $a_R \pm 3s$ has only a very low chance of bracketing the actual grade a_L whose determination is the object of the exercise. This underlines the importance—usually unappreciated—of what needs to be done in order

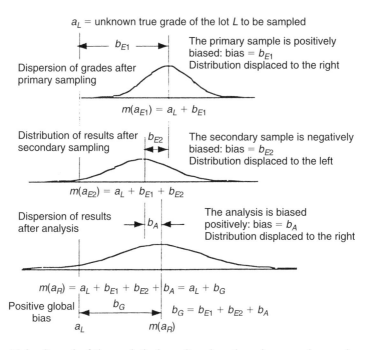

Figure A1.3 Spread of the analytical results when the primary and secondary sample preparation stages and the analysis are all biased

to eliminate all sources of bias from the primary and secondary sampling stages and from the analytical process too.

Estimators and estimates

A distinction needs to be drawn, one that is neither evident nor universally recognised, which warrants an explanation. The author has mentioned it in his publications over the past twenty years without it, apparently, raising any controversy. The preceding sections have provided illustrations of what is meant by an 'estimator'.

No one argues the fact that the analytical result a_R is an 'estimate' of the true but unknown grade a_{E2} of the final pulp taken for analysis. But can it be said that a_R is an 'estimate' of a_L? In the author's eyes, no! Two distinct operations have intervened in the passage from a_L to a_R: sampling and analysis.

The grade a_L of lot L cannot be estimated directly (see Section 1.2). First, lot L is substituted by the sample E_1, and this in its turn is substituted by the sample E_2, and it is the latter that is submitted for analysis (in practice there are more than two sampling stages, but for convenience they have been replaced by two major stages as was done in Chapter 1). The author takes the view that the true but unknown grade a_{E1} of the sample E_1 cannot be taken to be an estimate of a_L because it is not the result of a measure made on L. It is a quantity, used in place of a_L, that is not directly accessible. The preferred definition is that it is an 'estimator' of a_L.

In conclusion, for completeness and rigour a_R is an 'estimate' of the true unknown grade a_{E2} which itself is an 'estimator' of the true unknown grade a_{E1}, which in its turn is an 'estimator' of the true unknown grade a_L. It is easy enough to understand readers who consider that the analytical result a_R is ultimately a 'final estimate' of a_L; it is so much easier to accept! But one of the objects of this book is that readers should grasp the mechanics of sampling, and this requires that they understand the logical path that allows the value of a_R to be promoted to the value of a_L.

Appendix 2: The Three Branches of Statistics

THE NON-WEIGHTED STATISTICS OF POPULATIONS (FIGURE A2.1)

These could be called the 'statistics of populations of non-ordered objects F_i characterised by a single parameter'. This could be, for example, the value K_i of a certain parameter K. The objects are given, *a priori*, the same statistical weight.

A population L of N_L objects F_i characterised by the single parameter K_i is itself characterised by the mean parameter K_L such that:

$$K_L = \Sigma K_i / N_L \quad \text{with} \quad i = 1, 2, \ldots, N_L$$

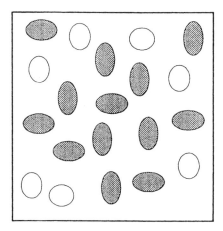

Figure A2.1   Diagrammatic representation of a population of objects with equal statistical weight

It applies to:

- Either material objects of equal or nearly-equal mass;
- Or immaterial objects of the same statistical weight such as certain analytical results or the opinions expressed in opinion polls.

This subject is dealt with in all the statistical text books and taught everywhere.

THE WEIGHTED STATISTICS OF POPULATIONS (FIGURE A2.2)

These could be called the 'statistics of non-ordered objects F characterised by two parameters'. The two parameters are:

1. A qualitative parameter such as the grade a in analyte A
2. A quantitative weighting parameter such as the actual, physical mass M of the object considered.

A population L of N_L objects F_i characterised by the parameter pair $(a_i M_i)$ is itself characterised by the parameter pair $(a_L M_L)$ that obeys the following rules:

$$M_L = \Sigma M_i \text{ and } a_L M_L = \Sigma a_i M_i \quad \text{with } i = 1, 2, \ldots, N_L$$

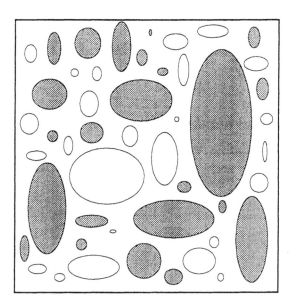

Figure A2.2 Diagrammatic representation of a population of objects with different statistical weights

It applies to:

• Either material objects of very different masses which must be allocated statistical weights proportional to their masses
• Or immaterial objects to which it is believed weighting factors are best applied, for example to analytical results for which a weighted mean is calculated.

This subject has been dealt with in the author's publications since 1951.

CHRONOSTATISTICS (FIGURE A2.3)

These could be called the 'statistics of ordered series of objects whose statistical weights are not necessarily equal'. It applies to:

• Either continuous material objects such as flowing streams of material
• Or discrete material objects such as the increments taken at uniform intervals from continuous material objects: or of manufactured objects leaving a plant or workshop; or again of contiguous or regularly spaced sections taken from a linear object of more or less constant cross-section (e.g. a steel rail).

'Autocorrelation' is the correlation that exists, at least potentially, between the units of the series. Autocorrelation is the cause of the principal difference between a series (chronological or spatial) and a population within which the idea of correlation between its units makes no sense. The tools used in chronostatistics are completely different from those used in classical statistics.

This subject has been treated in the author's publications since 1962.

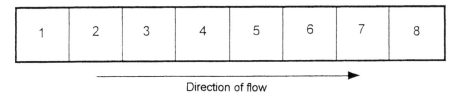

Direction of flow

Figure A2.3   Diagrammatic representation of a series of ordered objects

Appendix 3: Integral Calculus Nomenclature

The object of this appendix is to remove any confusion readers may have encountered on seeing the expressions appearing in Sections 12.3 and 12.4. When readers see an expression such as the following (e.g. Section 6.2.5)

$$a_L = \frac{1}{T_L} \int_0^{T_L} a(t)\, dt$$

they immediately understand that the function $a(t)$ in which t is the variable (the time) is integrated over the interval $[0, T_L]$ which extends from the times $t = 0$ to $t = T_L$. The values 0 and T_L are the limits of integration.

However, when they read in Section 12.4:

$$w'(j) = \frac{1}{j} \int_0^j w(j')\, dj' = \frac{2}{j^2} \int_0^j dj' \int_0^{j'} v(j'')\, dj''$$

unless they are familiar with the integral calculus they are right to question the significance and meanings of the quantities j, j' and j'' which denote, in fact, three quantities of the same kind. Comparing the following with the simple integral at the top of the page is it possible to write, without ambiguity

$$w'(j) = \frac{1}{j} \int_0^j w(j)\, dj?$$

Certainly not, because the same quantity j would play two different roles:

- In $w(j)$ j is a variable
- In the integral j is the upper limit of integration.

To avoid confusion, the current variable has been denoted by j'; its differential by dj'; and j has been reserved to describe the integration limit.

The same problem arises in the double integral. Taking the second (right-hand) integral: its variable is j'' and the limit of integration is j'. This second integral is a function of j'—in fact the area $S'(j)$ defined in Section 12.3—and then it becomes the object of a second integration in which j' is the variable and j is the integration limit.

The reader will see that it has been found necessary to use three different notations for the same quantity in order to avoid any mathematical ambiguity, as it plays three different roles in the same expression.

References

[1] Gy, P. 'Masse Minimale d'Echantillon Requise pour Representer un Lot de Minerai' (Minimum mass of sample needed to represent a mineral lot). *Norme interne a la Ste Minerais et Metaux* (unpublished) (1951).

[2] Gy, P. 'Erreur Commise dans le Prelevement d'un Echantillon sur un Lot de Minerai' (Error made when sampling a mineral lot). Congres des laveries des mines metalliques français, Paris (1953). *Rev. Ind. Minerale, St-Etienne*, **36**, 311–45 (1954).

[3] Roth, E. Avertissement aux lecteurs du chapiter P220 'Echantillonnage'. (The foreword to the chapter on sampling.) *Traité Analyse Chimique des Techniques de l'Ingenieur*, Paris, 2 (1989).

[4] Gy, P. 'Echantillonage'. Chapitre P220 du *Traité Analyse Chimique des Techniques de l'Ingenieur*, Paris, 1–20, (1989).

[5] Wegscheider, W. 'Richtige Probenahme Voraussetzung fur richtige Analysen' (Correct sampling: a pre-condition for correct analysis). Chapter 'Probenahme'. A set of papers published by Springer-Verlag, Vienna (1993).

[6] Gy, P. *Heterogeneity, Echantillonage, Homogeneisation. Ensemble Coherent de Theories* (*Heterogeneity, Sampling, Homogenisation. Their logical integration*). Masson, Paris (1988).

[7] Gy, P. *Sampling of Heterogeneous and Dynamic Material Systems. Theories of Heterogeneity, Sampling and Homogenising.* Elsevier, Amsterdam (1992).

[8] Pretsch, E. 'The Forgotten Bias'. Guest Editorial, *TrAC*, **14**, 2 (1995).

[9] Gy, P. 'Introduction to the Theory of Sampling. Part 1: Heterogeneity of a Population of Uncorrelated Units'. *TrAC*, **14**, 67–76 (1995).

[10] Brunton, D. W. 'The Theory and Practice of Ore Sampling'. *Trans. AIME*, **25**, 836 (1895).

[11] Ng, Kin C. *et al.* 'Digital Chemical Analysis of dilute microdroplets'. *Anal. Chem.*, **64**, 2914–19 (1992).

[12] Soper, S. A. 'Photon Burst Detection of Single Near-Infrared Fluorescent Molecules'. *Anal. Chem.*, **65**, 740–47 (1993).

[13] Pritchard, F. E. *et al. Quality in the Analytical Chemical Laboratory.* Chapter 2, 'Sampling' (1995).

[14] Matheron, G. *Les Variables Régionalisées et leur Estimation (Regionalised Variables and their Estimation).* Doctoral thesis. Masson, Paris (1965).

[15] David, M. *Handbook of Applied Advanced Geostatistical Ore Reserve Estimation.* Elsevier Science Publishers, Amsterdam (1988).

[16] Gy, P. 'L'Echantillonnage des Minerais en Vrac. Tome 1: Théorie Generale.' *Rev.*

Ind. Miner., Special Issue (15 January 1967). (The Sampling of Minerals in Bulk, Vol. 1, General Theory.)

[17] Gy, P. 'L'Echantillonage des Minerais en Vrac. Tome 2: Erreurs Operatoires.' *Rev. Ind. Miner.*, Special Issue (15 September 1971). (Vol. 2. Operational Errors.)

[18] Colijn, H. *Weighing and Proportioning of Bulk Solids.* Trans-Tech Publishers, Clausthal (1975).

[19] Gy, P. *Sampling of Particulate Materials, Theory and Practice*, Second revised edition. Elsevier, Amsterdam (1982).

Index